NATIONAL STANDARD OF THE PEOPLE'S REPUBLIC OF CHINA

Design Standard for Energy Efficiency of Public Buildings

GB 50189 – 2015

Chief Development Department:Ministry of Housing and Urban-Rural Development
of the People's Republic of China
Approval Department:Ministry of Housing and Urban-Rural Development
of the People's Republic of China
Implementation Date:October 1, 2015

China Architecture & Building Press

Beijing 2015

图书在版编目(CIP)数据

公共建筑节能设计标准 GB 50189-2015/中华人民共和国住房和城乡建设部组织编译. —北京：中国建筑工业出版社，2019.6
(工程建设标准英文版)
ISBN 978-7-112-23589-6

Ⅰ.①公… Ⅱ.①中… Ⅲ.①公共建筑-节能-建筑设计-设计标准-中国-英文 Ⅳ.①TU242-65

中国版本图书馆 CIP 数据核字(2019)第 068155 号

Chinese edition first published in the People's Republic of China in 2015

English edition first published in the People's Republic of China in 2019

by China Architecture & Building Press

No. 9 Sanlihe Road

Beijing, 100037

www.cabp.com.cn

Printed in China by Beijing YanLinJiZhao Printing Co., LTD

© 2015 by Ministry of Housing and Urban-Rural Development of
the People's Republic of China

All rights reserved. No part of this publication may be reproduced or transmitted in any form or by any means, graphic, electronic, or mechanical, including photocopying, recording, or any information storage and retrieval systems, without written permission of the publisher.

This book is sold subject to the condition that it shall not, by way of trade or otherwise, be lent, re-sold, hired out or otherwise circulated without the publisher's prior consent in any form of blinding or cover other than that in which this is published and without a similar condition including this condition being imposed on the subsequent purchaser.

ISBN 978-7-112-23589-6(33873)

Announcement of the Ministry of Housing and Urban-Rural Development of the People's Republic of China

No. 739

Announcement of the Ministry of Housing and Urban-Rural Development on Publishing the National Standard *Design Standard for Energy Efficiency of Public Buildings*

Design Standard for Energy Efficiency of Public Buildings has been approved as a national standard with a serial number of GB 50189-2015 and shall be implemented from October 1, 2015. Thereinto, articles 3.2.1, 3.2.7, 3.3.1, 3.3.2, 3.3.7, 4.1.1, 4.2.2, 4.2.3, 4.2.5, 4.2.8, 4.2.10, 4.2.14, 4.2.17, 4.2.19, 4.5.2, 4.5.4 and 4.5.6 are mandatory provisions and must be enforced strictly. Former *Design Standard for Energy Efficiency of Public Buildings* (GB 50189-2005) shall be abolished simultaneously.

Authorized by Research Institute of Standards & Norms, this standard is published and distributed by China Architecture & Building Press.

Ministry of Housing and Urban-Rural Development of the People's Republic of China
February 2, 2015

Foreword

According to the requirements of "Notice on Printing Development and Revision Plan of National Engineering Construction Standards and Codes in 2012" (JIANBIAO [2012] No. 5) issued by Ministry of Housing and Urban-Rural Development of the people's Republic of China, this standard is revised by the standard drafting group on the basis of extensive investigation and careful summarization of practical experience and by referring to the relevant international and foreign advanced standards and widely soliciting for opinions.

Main technical contents of this standard: 1. General Provisions; 2. Terms; 3. Building and Envelope Thermal Design; 4. Heating, Ventilation and Air Conditioning; 5. Water Supply and Drainage; 6. Electric; 7. Renewable Energy Application.

Main technical contents revised in this standard are: 1. setting up a typical public building model database representing characteristics and distribution characteristics of public buildings in China, on such basis determining the goal of energy efficiency of this standard; 2. updating building envelope thermal performance limits and cooling source energy efficiency limits, and making requirements respectively according to building classification and building thermal zones; 3. adding precondition for building envelope trade-off, supplementing and subdividing the requirements of trade-off calculation software and input and output content; 4. adding relevant requirements of water supply and drainage system, electrical system and renewable energy application.

The provisions printed in bold type in this standard are compulsory and must be enforced strictly.

Ministry of Housing and Urban-Rural Development is in charge of the administration of this standard and the explanation of the compulsory provisions. China Academy of Building Research is responsible for the explanation of specific technical contents. If there is any comment or suggestion during the process of implementing this standard, please send it to the drafting group for *Design Standard for Energy Efficiency of Public Buildings* of China Academy of Building Research (address: No. 30, North 3rd Ring East Road, 100013, Beijing).

Chief Development Organization:
  China Academy of Building Research

Participating Development Organizations:
  Beijing Institute of Architectural Design
  China Architecture Design and Research Group
  Shanghai Construction Design and Research Institute Co., Ltd.
  China Southwest Architectural Design & Research Institute Co., Ltd.
  Tianjin Architecture Design Institute
  Tongji University Architectural Design (Group) Co., Ltd. (TJAD)
  Northwestern Architectural Design Institute Co., Ltd.
  China Northeast Architectural Design & Research Institute Co., Ltd.

Sino-German College Applied Sciences of Tongji University
Shenzhen Institute of Building Research
Shanghai Academy of Building Research
Xinjiang Architecture Design & Research Institute
China Construction Design International (CCDI)
Shandong Provincial Architectural Design Institute
Central-South Architectural Design Institute Co., Ltd.
Architect Design & Research Institute of South China University of Technology
Zhongkai University of Agriculture and Engineering
Technavotor International Limited
Carrier Air Conditioning Sales & Service (Shanghai) Co., Ltd.
Trane Air Conditioning (China) Co., Ltd.
Daikin (China) Investment Co., Ltd.
Johnson Controls Building Equipment Technology (Wuxi) Co., Ltd.
Beijing JEG New Energy Tech. Development Co., Ltd.
Beijing Siemens Cerberus Electronics Ltd. (BSCE)
Beijing GBSWARE Software Co., Ltd.
Gree Electric Appliances, Inc. of Zhuhai
Shenzhen Fangda Building Technology Group Co., Ltd.
Owens Corning (China) Investment Co., Ltd.
MENRED Group
Guangdong AKE Technology Co., Ltd.
Orient Sundar Windoor Group
Beijing Zhenli Energy Conservation and Environmental Protection Technology Co., Ltd.

Chief Drafting Staff:

Xu Wei Zou Yu Xu Hongqing Wan Shuie
Pan Yungang Shou Weiwei Chen Qi Xu Feng
Feng Ya Gu Fang Che Xueya Liu Peng
Wang Qian Jin Lina Long Weiding Zhao Xiaoyu
Liu Mingming Liu Ming Mao Hongwei Zhou Hui
Yu Xiaoming Ma Youcai Chen Zuming Ding Lixing
Liu Junyue Chen Xi Sun Deyu Yang Liming
Shi Minqi Zhong Ming Shi Wen Ban Guangsheng
Shao Kangwen Liu Qiyao Chen Jin Zeng Xiaowu
Tian Hui Chen Linan Li Feilong Wei Hedong
Huang Zhenli Wang Biling Liu Zongjiang

Chief Examiners of This Standard:

Lang Siwei Sun Minsheng Jin Hongxiang Xu Huadong
Zhao Li Dai Deci Wu Xueling Zhang Xu
Zhao Shihuai Zhi Jianmin Wang Suying

Contents

1 General Provisions ··· (1)
2 Terms ·· (2)
3 Building and Envelope Thermal Design ·· (4)
　3.1　General Requirements ·· (4)
　3.2　Architectural Design ··· (5)
　3.3　Building Envelope Thermal Design ·· (7)
　3.4　Building Envelope Thermal Performance Trade-off ·························· (12)
4 Heating, Ventilation and Air Conditioning ··· (14)
　4.1　General Requirements ·· (14)
　4.2　Heating and Cooling Source ··· (15)
　4.3　Transmission and Distribution System ·· (21)
　4.4　Terminal System ··· (28)
　4.5　Monitor, Control and Measure ·· (28)
5 Water Supply and Drainage ··· (31)
　5.1　General Requirements ·· (31)
　5.2　Water Supply and Drainage System ··· (31)
　5.3　Service Water Heating ·· (32)
6 Electric ·· (34)
　6.1　General Requirements ·· (34)
　6.2　Power Supply and Distribution System ··· (34)
　6.3　Lighting ·· (34)
　6.4　Electric Power Supervision and Measure ······································· (35)
7 Renewable Energy Application ·· (37)
　7.1　General Requirements ·· (37)
　7.2　Solar Energy Application ··· (37)
　7.3　Ground Source Heat Pump System ··· (38)
Appendix A　Calculation of Mean Heat Transfer Coefficient of External Walls ············· (39)
Appendix B　Building Envelope Thermal Performance Trade-off ··························· (40)
Appendix C　Building Envelope Thermal Performance Compliance Form ················ (46)
Appendix D　Insulation Thickness of Pipes, Ducts and Equipments ······················· (48)
Explanation of Wording in This Standard ··· (51)
List of Quoted Standards ·· (52)

1 General Provisions

1.0.1 This standard is formulated with a view to implementing the relevant national laws and regulations and principles and policies, improving indoor environment of public buildings, increasing energy utilization efficiency, promoting renewable energy application in buildings, and reducing building energy consumption.

1.0.2 This standard is applicable to energy efficiency design for new, extended and renovated public buildings.

1.0.3 Energy efficiency design for public buildings shall, according to local climate conditions, improve building envelope insulation, increase energy utilization efficiency of building equipment and system, and reduce energy consumption by heating, ventilation and air conditioning, water supply and drainage and electrical system of buildings using renewable energy under the premises of ensuring indoor environment parameters.

1.0.4 When the building height is more than 150m or the above-ground building area of a single building is more than 200,000m^2, not only shall the requirements of this standard be complied with, but also experts shall be organized to conduct special demonstration on the energy efficiency design.

1.0.5 The energy-efficient measures for the project shall be indicated in the construction drawing design document. Besides, the operating requirements are suggested to be described.

1.0.6 Energy efficiency design for public buildings shall not only meet the requirements of this standard, but also comply with those in the current relevant ones of the nation.

2 Terms

2.0.1 Transparent curtain wall

The curtain wall through which visible light may be directly transmitted into the room.

2.0.2 Shape factor

The ratio of the building external surface area contacting with the outdoor atmospheric air to the enclosed volume of the building. The external surface area excludes the area of ground, and non-heating staircase internal walls.

2.0.3 Single facade window to wall ratio

The ratio of the window opening area on a facade of building to the total area of the facade, window to wall ratio for short.

2.0.4 Solar heat gain coefficient ($SHGC$)

The ratio of solar radiation obtained in the room through transparent building envelope (door and window or transparent curtain wall) to solar radiation projected on the external surface of transparent building envelope (door and window or transparent curtain wall). The amount of solar radiation obtained in the room includes heat obtained from solar radiation through radiation transmission and that of solar radiation transmitted into the room after absorbed by members.

2.0.5 Visible transmittance

The ratio of luminous flux of the visible light through transparent materials to that of the visible light projected on their surfaces.

2.0.6 Building envelope thermal performance trade-off

It refers to a method to calculate and compare full-year energy consumption by heating and air conditioning of reference building and designed building, and to judge if the overall thermal performance of building envelope meets the requirements of energy efficiency design, when architectural design fails to fully meet the index requirements specified in the Building Envelope Thermal Design, trade-off for short.

2.0.7 Reference building

The benchmark building used to calculate full-year energy consumption by heating and air conditioning for compliance with the standard requirements, in building envelope thermal performance trade-off.

2.0.8 Integrated part load value ($IPLV$)

A single value, based on the performance coefficient value of the unit under part load, weighted according to the percentage of accumulated load of the unit under various loading conditions, to calculate the efficiency of water chilling unit at part load for air conditioning.

2.0.9 Electricity consumption to transferred heat quantity ratio ($EHR\text{-}h$)

The ratio of total power consumption (kW) of circulating water pump in central heating system to design thermal load (kW) under design conditions.

2.0.10 Electricity consumption to transferred cooling (heat) quantity ratio [$EC(H)R\text{-}a$]

The ratio of total power consumption (kW) of circulating water pump in air conditioning chilled/hot water system to designed cooling (heating) load under design conditions.

2.0.11 System coefficient of refrigeration performance (*SCOP*)

The ratio of refrigerating capacity of electrically driven refrigerating system to net energy input of refrigerator, cooling water pump and cooling tower under design conditions.

2.0.12 Energy consumption per unit air volume of air duct system (W_s)

Electric power (W) consumed by air duct system for air conditioning and ventilation by transmitting per unit air volume (m³/h) under design conditions.

3 Building and Envelope Thermal Design

3.1 General Requirements

3.1.1 Public buildings shall be classified according to the following requirements:

1 The buildings with single building area more than 300 m² or the building complex with single building area less than or equal to 300m² but total building area more than 1,000m² shall be Category A public buildings;

2 The buildings with single building area less than or equal to 300m² shall be Category B public buildings.

3.1.2 Building thermal design zones of representative cities shall be determined according to Table 3.1.2.

Table 3.1.2 Building Thermal Design Zone of Representative Cities

Climate zones and climate sub-zones		Representative cities
Severe cold zone	Severe cold zone A	Boketu, Yichun, Huma, Hailar, Manzhouli, Arxan, Madoi, Heihe, Nen River, Hailun, Qiqihar, Fujin, Harbin, Mudanjiang, Daqing, Anda, Jiamusi, Erenhot, Duolun, Da Qaidam, Altay, Naqu
	Severe cold zone B	
	Severe cold zone C	Changchun, Tonghua, Yanji, Tongliao, Siping, Fushun, Fuxin, Shenyang, Benxi, Anshan, Hohhot, Baotou, Erdos, Chifeng, Ejina County, Datong, Urumqi, Karamay, Jiuquan, Xining, Shigatse, Ganzi, Kangding
Cold zone	Cold zone A	Dandong, Dalian, Zhangjiakou, Chengde, Tangshan, Qingdao, Luoyang, Taiyuan, Yangquan, Jincheng, Tianshui, Yulin, Yan'an, Baoji, Yinchuan, Pingliang, Lanzhou, Kashgar, Yining, Aba, Lhasa, Nyingchi, Beijing, Tianjin, Shijiazhuang, Baoding, Xingtai, Jinan, Dezhou, Yanzhou, Zhengzhou, Anyang, Xuzhou, Yuncheng, Xi'an, Xianyang, Turpan, Korla, Hami
	Cold zone B	
Hot summer and cold winter zone	Hot summer and cold winter zone A	Nanjing, Bengbu, Yancheng, Nantong, Hefei, Anqing, Jiujiang, Wuhan, Huangshi, Yueyang, Hanzhong, Ankang, Shanghai, Hangzhou, Ningbo, Wenzhou, Yichang, Changsha, Nanchang, Zhuzhou, Yongzhou, Ganzhou, Shaoguan, Guilin, Chongqing, Dachuan, Wanzhou, Fuling, Nanchong, Yibin, Chengdu, Zunyi, Kaili, Mianyang, Nanping
	Hot summer and cold winter zone B	
Hot summer and warm winter zone	Hot summer and warm winter zone A	Fuzhou, Putian, Longyan, Meizhou, Xingning, Yingde, Hedi, Liuzhou, Hezhou, Quanzhou, Xiamen, Guangzhou, Shenzhen, Zhanjiang, Shantou, Nanning, Beihai, Wuzhou, Haikou, Sanya
	Hot summer and warm winter zone B	
Temperate zone	Temperate zone A	Kunming, Guiyang, Lijiang, Huize, Tengchong, Baoshan, Dali, Chuxiong, Qujing, Luxi, Pingbian, Guangnan, Xingyi, Dushan
	Temperate zone B	Ruili, Gengma, Lincang, Lancang, Simao, Jiangcheng, Mengzi

3.1.3 Relief of heat island effect shall be taken into consideration in the general planning of the architectural complex. General planning and general layout design of buildings shall facilitate to natural ventilation and sunlight in winter. The local optimal orientation or optimum orientation is suggested to be selected as the main building orientation, and the buildings is should be kept clear from the prevailing wind direction in winter.

3.1.4 Architectural design shall follow the priority principle of passive energy-efficient measures and reduce energy demand of buildings by making the best of natural lighting and ventilation in combination with thermal insulation and sun shading measures of building envelope.

3.1.5 Buildings are suggested to be shaped regularly and compactly, avoiding excessive change of concavo-concave.

3.1.6 In the general layout design and plane layout of buildings, it is required to reasonably determine position of energy equipment room and shorten energy supply conveying distance. Heating/cooling source equipment rooms of the same public building are suggested to be arranged intensively at or close to the central position of cooling/thermal load.

3.2 Architectural Design

3.2.1 Shape factor of public buildings in severe cold and cold zone must be in accordance with those specified in Table 3.2.1.

Table 3.2.1 Shape factor of Public Buildings in Severe Cold and Cold zone

Single building area A (m^2)	Shape factor
300<A≤800	≤0.50
A>800	≤0.40

3.2.2 Single facade window to wall ratio (including transparent curtain wall) of Category A public buildings in severe cold zone is not suggested to be greater than 0.60; that in other zones is not suggested to be greater than 0.70 (including transparent curtain wall).

3.2.3 Single facade window to wall ratio shall be calculated according to the following requirements:

  1 Orientation of concavo-convex facade shall be calculated according to the orientation of the facade it is located;

  2 External wall and external window of staircase and elevator room shall be counted in;

  3 Area of top, bottom and sidewall of bay window shall not be counted in the external wall area;

  4 When the external window, top and side on the external wall is bay window with opaque, window area shall be calculated according to the window opening area; when the top and side of bay window is transparent, bay window area shall be calculated according to the actual transparent area.

3.2.4 When single facade window to wall ratio of Category A public buildings is less than 0.40, visible transmittance of transparent materials shall not be less than 0.60; When single facade window to wall ratio of Category A public buildings is greater than or equal to 0.40, visible transmittance of transparent materials shall not be less than 0.40.

3.2.5 Sun shading measures shall be taken for the external windows (including transparent curtain wall) in all orientation of the buildings in hot summer and warm winter zones, hot summer and cold winter zones and temperate zones; sun shading measure is suggested to be taken for the buildings in cold zones. Arrangement of external sunshade is suggested to meet the following requirements:

    1 Movable external sunshade is suggested to be arranged in the east-west direction and horizontal external sunshade is suggested to be arranged in the south direction;

    2 External sun shading devices of buildings shall give consideration to ventilation and sunlight in winter.

3.2.6 Classification of building facade orientation shall meet the following requirements:

    1 In the north, north by west 60° to north by east 60°;

    2 In the south, south by west 30° to south by east 30°;

    3 In the west, west by north 30° (including) to west by south 60° (including);

    4 In the east, east by north 30° (including) to east by south 60° (including);

3.2.7 Transparent area on the roof of Category A public buildings must not be greater than 20% of the total roof area. Otherwise, trade-off must be carried out according to the method specified in this standard.

3.2.8 Effective ventilation and air change area of single facade external window (including transparent curtain wall) shall meet the following requirements:

    1 Operable sash shall be arranged on the external window (including transparent curtain wall) of Category A public buildings, and its effective ventilation and air change area is not suggested to be less than 10% of the external wall area of the room; ventilation and air change equipment shall be arranged if it is impossible to arrange operable sash due to restrictions of conditions for transparent curtain wall.

    2 Effective ventilationand air change area of external window of Category B public buildings is not suggested to be less than 30% of the window area.

3.2.9 Effective ventilation and air change area of external window (including transparent curtain wall) shall be area of operable sash or ventilation interface area when window is open, whichever is smaller.

3.2.10 Storm porch shall be arranged for the external doors of buildings in severe cold zone; for the buildings in cold zone, Storm porch or double-layer external doors shall be arranged for the external doors toward the prevailing wind direction, Storm porch or other measures isolating cold blast is suggested to be arranged for other external doors; as for buildings in hot summer and cold winter zone, hot summer and warm winter zone and temperate zone, thermal insulation measures shall be taken for the external doors.

3.2.11 Building atrium shall take full advantage of natural ventilation for cooling, where mechanical air exhaust device may be arranged to strengthen natural air supply.

3.2.12 Building design shall take full advantage of natural daylight. In the locations where natural daylight fails to meet the illumination requirements, light guiding, reflecting and other devices are suggested to be adopted to introduce natural daylight into the rooms.

3.2.13 Visible light reflectance of the internal surfaces in the rooms where people stay for long is suggested to meet the requirements specified in Table 3.2.13.

Table 3.2.13 Visible Light Reflectance of the Internal Surfaces in the Rooms where People Stay for Long

Locations of surfaces in the rooms	Visible light reflectance
Ceiling	0.7~0.9
Wall surface	0.5~0.8
Ground surface	0.3~0.5

3.2.14 Elevators shall have the functions of energy efficiency operation. Group control measures shall be taken where two or more elevators are arranged centralized. Elevators shall have the functions of automatic switch to energy efficiency operation mode if there is no external call and no preset command in the elevator car within a period of time.

3.2.15 The escalator and moving sidewalk shall have the functions of unloaded suspend and slow speed operation.

3.3 Building Envelope Thermal Design

3.3.1 According to climate zones of building thermal design, building envelope thermal performance of Category A public buildings must meet the requirements of Tables 3.3.1.1~3.3.1-6 respectively. Trade-off must be carried out according to the method specified in this standard if it fails to meet the requirements of this article.

Table 3.3.1-1 Limits of Building Envelope Thermal Performance of Category A Public Buildings in Severe Cold Zones A and B

Building envelope parts		Shape factor ≤0.30	0.30<Shape factor≤0.50
		Heat transfer coefficient K [W/(m² · K)]	
Roofing		≤0.28	≤0.25
External wall (including non-transparent curtain wall)		≤0.38	≤0.35
Overhead or overhanging floor-slab with bottom surface contacting outdoor air		≤0.38	≤0.35
Floor slab between underground garage and heating room		≤0.50	≤0.50
Partition wall between non-heating staircase and heating room		≤1.2	≤1.2
Single facade external window (including transparent curtain wall)	Window to wall ratio≤0.20	≤2.7	≤2.5
	0.20<Window to wall ratio≤0.30	≤2.5	≤2.3
	0.30<Window to wall ratio≤0.40	≤2.2	≤2.0
	0.40<Window to wall ratio≤0.50	≤1.9	≤1.7
	0.50<Window to wall ratio≤0.60	≤1.6	≤1.4
	0.60<Window to wall ratio≤0.70	≤1.5	≤1.4
	0.70<Window to wall ratio≤0.80	≤1.4	≤1.3
	Window to wall ratio>0.80	≤1.3	≤1.2
Transparent part on the roof (transparent area on the roof ≤20%)		≤2.2	
Building envelope parts		Thermal resistance of insulation material layer R [(m² · K)/W]	
Surrounding ground		≥1.1	
External wall of heating basement in contact with soil		≥1.1	
Deformation joint (in the case of internal insulation of two side walls)		≥1.2	

Table 3.3.1-2 Limits of Building Envelope Thermal Performance
of Category A Public Buildings in Severe Cold Zone C

Building envelope parts		Shape factor≤0.30	0.30<Shape factor≤0.50
		Heat transfer coefficient K [W/(m²·K)]	
Roofing		≤0.35	≤0.28
External wall (including non-transparent curtain wall)		≤0.43	≤0.38
Overhead or overhanging floor-slab with bottom surface contacting outdoor air		≤0.43	≤0.38
Floor slab between underground garage and heating room		≤0.70	≤0.70
Partition wall between non-heating staircase and heating room		≤1.5	≤1.5
Single facade external window (including transparent curtain wall)	Window to wall ratio≤0.20	≤2.9	≤2.7
	0.20<Window to wall ratio≤0.30	≤2.6	≤2.4
	0.30<Window to wall ratio≤0.40	≤2.3	≤2.1
	0.40<Window to wall ratio≤0.50	≤2.0	≤1.7
	0.50<Window to wall ratio≤0.60	≤1.7	≤1.5
	0.60<Window to wall ratio≤0.70	≤1.7	≤1.5
	0.70<Window to wall ratio≤0.80	≤1.5	≤1.4
	Window to wall ratio>0.80	≤1.4	≤1.3
Transparent part on the roof (transparent area on the roof ≤20%)		≤2.3	
Building envelope parts		Thermal resistance of insulation material layer R [(m²·K)/W]	
Surrounding ground		≥1.1	
External wall of heating basement in contact with soil		≥1.1	
Deformation joint (in the case of internal insulation of two side walls)		≥1.2	

Table 3.3.1-3 Limits of Building Envelope Thermal Performance of Category A Public Buildings in Cold Zone

Building envelope parts		Shape factor≤0.30		0.30<Shape factor≤0.50	
		Heat transfer coefficient K [W/(m²·K)]	Solar heat gain coefficient SHGC (east, south, west/north)	Heat transfer coefficient K [W/(m²·K)]	Solar heat gain coefficient SHGC (east, south, west/north)
Roofing		≤0.45	—	≤0.40	—
External wall (including non-transparent curtain wall)		≤0.50	—	≤0.45	—
Overhead or overhanging floor-slab with bottom surface contacting outdoor air		≤0.50	—	≤0.45	—
Floor slab between underground garage and heating room		≤1.0	—	≤1.0	—
Partition wall between non-heating staircase and heating room		≤1.5	—	≤1.5	—
Single facade external window (including transparent curtain wall)	Window to wall ratio≤0.20	≤3.0	—	≤2.8	—
	0.20<Window to wall ratio≤0.30	≤2.7	≤0.52/—	≤2.5	≤0.52/—
	0.30<Window to wall ratio≤0.40	≤2.4	≤0.48/—	≤2.2	≤0.48/—
	0.40<Window to wall ratio≤0.50	≤2.2	≤0.43/—	≤1.9	≤0.43/—
	0.50<Window to wall ratio≤0.60	≤2.0	≤0.40/—	≤1.7	≤0.40/—
	0.60<Window to wall ratio≤0.70	≤1.9	≤0.35/0.60	≤1.7	≤0.35/0.60
	0.70<Window to wall ratio≤0.80	≤1.6	≤0.35/0.52	≤1.5	≤0.35/0.52
	Window to wall ratio>0.80	≤1.5	≤0.30/0.52	≤1.4	≤0.30/0.52

Table 3.3.1-3(continued)

Building envelope parts	Shape factor≤0.30		0.30<Shape factor≤0.50	
	Heat transfer coefficient $K[W/(m^2 \cdot K)]$	Solar heat gain coefficient SHGC (east, south, west/north)	Heat transfer coefficient $K[W/(m^2 \cdot K)]$	Solar heat gain coefficient SHGC (east, south, west/north)
Transparent part on the roof (transparent area on the roof ≤20%)	≤2.4	≤0.44	≤2.4	≤0.35
Building envelope parts	Thermal resistance of insulation material layer $R[(m^2 \cdot K)/W]$			
Surrounding ground	≥0.60			
External wall of heating basement with air conditioner (wall in contact with soil)	≥0.60			
Deformation joint (in the case of internal insulation of two side walls)	≥0.90			

Table 3.3.1-4 Limits of Building Envelope Thermal Performance of Category A Public Buildings in Hot Summer and Cold Winter Zone

Building envelope parts		Heat transfer coefficient $K[W/(m^2 \cdot K)]$	Solar heat gain coefficient SHGC (east, south, west/north)
Roofing	Thermal inertia indexes of building envelope D≤2.5	≤0.40	—
	Thermal inertia indexes of building envelope D>2.5	≤0.50	
External wall (including non-transparent curtain wall)	Thermal inertia indexes of building envelope D≤2.5	≤0.60	—
	Thermal inertia indexes of building envelope D>2.5	≤0.80	
Overhead or overhanging floor-slab with bottom surface contacting outdoor air		≤0.70	—
Single facade external window (including transparent curtain wall)	Window to wall ratio≤0.20	≤3.5	—
	0.20<Window to wall ratio≤0.30	≤3.0	≤0.44/0.48
	0.30<Window to wall ratio≤0.40	≤2.6	≤0.40/0.44
	0.40<Window to wall ratio≤0.50	≤2.4	≤0.35/0.40
	0.50<Window to wall ratio≤0.60	≤2.2	≤0.35/0.40
	0.60<Window to wall ratio≤0.70	≤2.2	≤0.30/0.35
	0.70<Window to wall ratio≤0.80	≤2.0	≤0.26/0.35
	Window to wall ratio>0.80	≤1.8	≤0.24/0.30
Transparent part on the roof (transparent area on the roof ≤20%)		≤2.6	≤0.30

Table 3.3.1-5 Limits of Building Envelope Thermal Performance of Category A Public Buildings in Hot Summer and Warm Winter Zone

Building envelope parts		Heat transfer coefficient $K[W/(m^2 \cdot K)]$	Solar heat gain coefficient SHGC (east, south, west/north)
Roofing	Thermal inertia indexes of building envelope D≤2.5	≤0.50	—
	Thermal inertia indexes of building envelope D>2.5	≤0.80	

Table 3.3.1-5(continued)

Building envelope parts		Heat transfer coefficient K [W/(m²·K)]	Solar heat gain coefficient SHGC (east, south, west/north)
External wall (including non-transparent curtain wall)	Thermal inertia indexes of building envelope D≤2.5	≤0.80	—
	Thermal inertia indexes of building envelope D>2.5	≤1.5	
Overhead or overhanging floor-slab with bottom surface contacting outdoor air		≤1.5	—
Single facade external window (including transparent curtain wall)	Window to wall ratio≤0.20	≤5.2	≤0.52/—
	0.20<Window to wall ratio≤0.30	≤4.0	≤0.44/0.52
	0.30<Window to wall ratio≤0.40	≤3.0	≤0.35/0.44
	0.40<Window to wall ratio≤0.50	≤2.7	≤0.35/0.40
	0.50<Window to wall ratio≤0.60	≤2.5	≤0.26/0.35
	0.60<Window to wall ratio≤0.70	≤2.5	≤0.24/0.30
	0.70<Window to wall ratio≤0.80	≤2.5	≤0.22/0.26
	Window to wall ratio>0.80	≤2.0	≤0.18/0.26
Transparent part on the roof (transparent area on the roof ≤20%)		≤3.0	≤0.30

Table 3.3.1-6 Limits of Building Envelope Thermal Performance of Category A Public Buildings in Temperate Zone

Building envelope parts		Heat transfer coefficient K [W/(m²·K)]	Solar heat gain coefficient SHGC (east, south, west/north)
Roofing	Thermal inertia indexes of building envelope D≤2.5	≤0.50	
	Thermal inertia indexes of building envelope D>2.5	≤0.80	
External wall (including non-transparent curtain wall)	Thermal inertia indexes of building envelope D≤2.5	≤0.80	
	Thermal inertia indexes of building envelope D>2.5	≤1.5	
Single facade external window (including transparent curtain wall)	Window to wall ratio≤0.20	≤5.2	—
	0.20<Window to wall ratio≤0.30	≤4.0	≤0.44/0.48
	0.30<Window to wall ratio≤0.40	≤3.0	≤0.40/0.44
	0.40<Window to wall ratio≤0.50	≤2.7	≤0.35/0.40
	0.50<Window to wall ratio≤0.60	≤2.5	≤0.35/0.40
	0.60<Window to wall ratio≤0.70	≤2.5	≤0.30/0.35
	0.70<Window to wall ratio≤0.80	≤2.5	≤0.26/0.35
	Window to wall ratio>0.80	≤2.0	≤0.24/0.30
Transparent part on the roof (transparent area on the roof ≤20%)		≤3.0	≤0.30

Note: heat transfer coefficient (K) is only applicable to Temperate Zone A; heat transfer coefficient (K) of Temperate Zone B is not required.

3.3.2 Building envelope thermal performance of Category B public buildings must be in accordance with those specified in Tables 3.3.2-1 and 3.3.2-2.

Table 3.3.2-1 Limits of Thermal Performance for Roofing, External
Walls and Floor Slabs of Category B Public Buildings

Building envelope parts	Heat transfer coefficient K [W/(m² · K)]				
	Severe cold zone A and B	Severe cold zone C	Cold zone	Hot summer and cold winter zone	Hot summer and warm winter zone
Roofing	≤0.35	≤0.45	≤0.55	≤0.70	≤0.90
External wall (including non-transparent curtain wall)	≤0.45	≤0.50	≤0.60	≤1.0	≤1.5
Overhead or overhanging floor-slab with bottom surface contacting outdoor air	≤0.45	≤0.50	≤0.60	≤1.0	—
Floor slab between underground garage and heating room	≤0.50	≤0.70	≤1.0	—	—

Table 3.3.2-2 Limits of Thermal Performance for External Window
(Including Transparent Curtain Wall) of Category B Public Buildings

Building envelope parts	Heat transfer coefficient K [W/(m² · K)]					Solar heat gain coefficient SHGC		
External window (including transparent curtain wall)	Severe cold zone A and B	Severe cold zone C	Cold zone	Hot summer and cold winter zone	Hot summer and warm winter zone	Cold zone	Hot summer and cold winter zone	Hot summer and warm winter zone
Single facade external window (including transparent curtain wall)	≤2.0	≤2.2	≤2.5	≤3.0	≤4.0	—	≤0.52	≤0.48
Transparent part on the roof (transparent area on the roof ≤20%)	≤2.0	≤2.2	≤2.5	≤3.0	≤4.0	≤0.44	≤0.35	≤0.30

3.3.3 Thermal performance parameters of building envelope shall be calculated according to the following requirements:

1 The heat transfer coefficient of the external wall refers to the mean heat transfer coefficient calculated including structural thermal bridge, and the mean heat transfer coefficient shall be worked out according to the requirements of Appendix A in this standard.

2 Heat transfer coefficient of external window (including transparent curtain wall) shall be calculated according to the relevant requirements of current national standard of GB 50176 *Thermal Design Code for Civil Building*.

3 If external sun shading members are arranged, solar heat gain coefficient of external window (including transparent curtain wall) shall be the product of solar heat gain coefficient of external window itself (including transparent curtain wall) and shading coefficient of external sun shading members. Solar heat gain coefficient of external window itself (including transparent curtain wall) and shading coefficient of external sun shading members shall be calculated according to the relevant requirements of current national standard of GB 50176 *Thermal Design Code for*

Civil Building .

3.3.4 Internal surface temperature at thermal bridge of roofing, external wall and basement shall not be less than the dew point temperature of indoor air.

3.3.5 The air tightness grade of external doors and windows shall meet the requirements of article 4.1.2 in the national standard of GB/T 7106-2008 *Graduations and Test Methods of Air Permeability Watertightness Wind Load Resistance Performance for Building External Windows and Doors* and the following requirements:

1 Air tightness of external windows of the buildings in 10 stories or more shall not be less than Grade 7;

2 Air tightness of external windows of the buildings in less than 10 stories shall not be less than Grade 6;

3 Air tightness of external doors of the buildings in severe cold and cold zone shall not be less than Grade 4.

3.3.6 Air tightness of building curtain wall shall meet the requirements of article 5.1.3 in the national standard of GB/T 21086-2007 *Building Curtain Walls* and shall not be less than Grade 3.

3.3.7 When full glass curtain wall is adopted for the entrance lobbies of public buildings, the area of non-hollow glass in full glass curtain wall must not exceed 15% of transparent area (door, window and glass curtain wall) on the same facade. Besides, mean heat transfer coefficient must be weighted according to transparent area (including area of full glass curtain wall) on the same facade.

3.4 Building Envelope Thermal Performance Trade-off

3.4.1 Before building envelope thermal performance trade-off, thermal performance of designed building shall be verified; trade-off may be made only if the following basic requirements are met:

1 Basic requirements for the heat transfer coefficient of roofing shall be in accordance with those specified in Table 3.4.1-1.

Table 3.4.1-1 Basic Requirements for the Heat Transfer Coefficient of Roofing

Heat transfer coefficient K [$W/(m^2 \cdot K)$]	Severe cold zones A and B	Severe cold zone C	Cold zone	Hot summer and cold winter zone	Hot summer and warm winter zone
	≤0.35	≤0.45	≤0.55	≤0.70	≤0.90

2 Basic requirements for the heat transfer coefficient of external wall (including non-transparent curtain wall) shall be in accordance with those specified in Table 3.4.1-2.

Table 3.4.1-2 Basic Requirements for the Heat Transfer Coefficient of External Wall (Including Non-transparent Curtain Wall)

Heat transfer coefficient K [$W/(m^2 \cdot K)$]	Severe cold zones A and B	Severe cold zone C	Cold zone	Hot summer and cold winter zone	Hot summer and warm winter zone
	≤0.45	≤0.50	≤0.60	≤1.0	≤1.5

3 When single facade window to wall ratio is greater than or equal to 0.40, basic requirements for heat transfer coefficient and integrated solar heat gain coefficient of external window (including transparent curtain wall) shall be in accordance with those specified in Table 3.4.1-3.

Table 3.4.1-3 Basic Requirements for Heat Transfer Coefficient and Solar Heat Gain Coefficient of External Window (Including Transparent Curtain Wall)

Climate zones	Window to wall ratio	Heat transfer coefficient K [W/(m² · K)]	Solar heat gain coefficient $SHGC$
Severe cold zones A and B	0.40<Window to wall ratio≤0.60	≤2.5	—
	Window to wall ratio>0.60	≤2.2	
Severe cold zone C	0.40<Window to wall ratio≤0.60	≤2.6	—
	Window to wall ratio>0.60	≤2.3	
Cold zone	0.40<Window to wall ratio≤0.70	≤2.7	—
	Window to wall ratio>0.70	≤2.4	
Hot summer and cold winter zone	0.40<Window to wall ratio≤0.70	≤3.0	≤0.44
	Window to wall ratio>0.70	≤2.6	
Hot summer and warm winter zone	0.40<Window to wall ratio≤0.70	≤4.0	≤0.44
	Window to wall ratio>0.70	≤3.0	

3.4.2 As for trade-off of thermal performance of building envelope, year-round heating and air conditioning energy consumption by reference building under specified conditions is suggested to be first calculated, and then year-round heating and air conditioning energy consumption of the designed building under identical conditions is suggested to be calculated; if heating and air conditioning energy consumption of the designed building is less than or equal to that by reference building, the overall thermal performance of building envelope shall be judged as conforming to the energy efficiency requirements. If heating and air conditioning energy consumption of the designed building is greater than that by reference building, design parameters shall be adjusted and recalculated until heating and air conditioning energy consumption of the designed building less than that by reference building.

3.4.3 Shape, size, orientation, window to wall ratio, internal space division and use function of reference building shall be accordance with those of the designed building completely. When the transparent part area on the roof of designed building is greater than the requirements of article 3.2.7 in this standard, the transparent part area on the roof of reference building shall be scaled down to make it meet the requirements of article 3.2.7 in this standard.

3.4.4 Thermal performance parameter values of reference building envelope shall be taken according to the requirements of article 3.3.1 in this standard. Construction of external wall and roofing of reference building shall be in conformity with that of the designed building. If solar heat gain coefficient of external window (including transparent curtain wall) is not specified in article 3.3.1 of this standard, solar heat gain coefficient of external window (including transparent curtain wall) of reference building shall be in conformity with that of the designed building.

3.4.5 Building envelope thermal performance trade-off shall meet the requirements of Appendix B of this standard; corresponding original information and calculation results shall be provided according to Appendix C of this standard.

4 Heating, Ventilation and Air Conditioning

4.1 General Requirements

4.1.1 Heating load calculation and item-by-item hourly cooling load calculation must be carried out during construction documents design phase of Category A public buildings.

4.1.2 In severe cold zones A and B, hot water central heating system is suggested to be arranged for the public buildings, and hot air terminals is not suggested to be adopted as the only heating mode for the buildings equipped with air conditioning systems; as for the public buildings in severe cold zone C and cold zone, heating mode shall be determined through comprehensive technical and economic analysis and comparison in accordance with such factors as grade of building, number of days in the heating period, energy consumption and operating cost.

4.1.3 Selection of system cooling/heating medium temperature shall meet the relevant requirements of the current national standard of GB 50736 *Design Code for Heating Ventilation and Air Conditioning of Civil Buildings*. If economic and technology condition is rational, cooling medium temperature is suggested to be higher than common designed temperature and heating medium temperature be lower than common designed temperature.

4.1.4 If indoor surplus heat, surplus moisture or other pollutants can be discharged by ventilation, natural ventilation, mechanical ventilation or compound ventilation mode is suggested to be adopted.

4.1.5 Disperse air conditioners or systems is suggested to be adopted if one of the following conditions is met:

    1 Short cooling or heating time required all year round, or uneconomic to adopt central cooling or heating system;

    2 Conditioned rooms are scattered;

    3 Several rooms with different conditioned time and requirements in the buildings with central cooling and heating systems;

    4 The existing public buildings to be additionally equipped with air conditioning systems where it is difficult to arrange equipment room and pipes.

4.1.6 When the air conditioning system with temperature and humidity independent control is adopted, the following requirements shall be met:

    1 High temperature cooling source preparation mode and fresh air dehumidification mode shall be determined based on technical and economic analysis with consideration of climate characteristics;

    2 Application measures for natural cooling source and renewable energy all year around is suggested to be taken into consideration;

    3 Reheating air handling mode is not suggested to be adopted.

4.1.7 The air-conditioned zones with different use time shall not be put into the same all air system with fixed air rate. The air-conditioned zones with different requirements for temperature, humidity and so on shall not be put into the same air conditioning air system.

4.2 Heating and Cooling Source

4.2.1 Heating and cooling sources for heating and air conditioning shall be determined through comprehensive demonstration according to building scale, use, energy condition, structure and price at the construction site, and the relevant requirements of national energy conservation and emission reduction and environmental protection policies, which shall meet the following requirements:

1 Waste heat or industrial surplus heat should be adopted as heat source in the zones where waste heat or industrial surplus heat can be utilized. In the case that waste heat or industrial surplus heat is at higher temperature, absorption chiller is suggested to be adopted as the cooling source if proven reasonable through technical and economic demonstration.

2 If economic and technology condition is rational, shallow geothermal energy, solar energy, wind energy and other renewable energy is suggested to be used as heating and cooling source. Auxiliary heating and cooling sources shall be arranged if adoption of renewable energy cannot be guaranteed all year round due to restrictions of such factors as climate.

3 In the regions not meeting the conditions in items 1 and 2 of this article but with urban or regional heating network, heating source of centralized air conditioning system is suggested to give priority to the urban or regional heating network.

4 In the regions not meeting the conditions in items 1 and 2 of this article but with sufficient power supply from urban power grid in summer, electric compression-type unit is suggested to be adopted as cooling source of air conditioning system.

5 In the regions not meeting the conditions in items 1~4 of this article but with sufficient urban gas supply, gas boiler and gas heater are suggested to be adopted for heating or gas absorption chiller-heater for cooling or heating.

6 In the regions not meeting the conditions in items 1 ~ 5 of this article, coal-fired boiler and oil-fired boiler may be adopted for heating, steam absorption chiller or fuel absorption chiller-heater for cooling or heating.

7 In the regions with lower designed dew point temperature of outdoor air in summer, the water chilling unit that cools by indirect evaporation is suggested to be adopted as the cooling source of air conditioning system.

8 In the regions with sufficient natural gas supply, if power load, heating load and cooling load of buildings can be matched preferably, and comprehensive energy utilization efficiency of combined heating cooling and power production system can be maximized with proven reasonable through economic and technical comparison, distributed gas combined cooling heating and power system should be adopted.

9 For the buildings requiring simultaneous heating and cooling for long and air conditioning all year around with great difference ofeach room or each area load characteristics, water-loop heat pump air conditioning system is suggested to be adopted for cooling and heating if the technical and economical comparison is reasonable.

10 In the regions where time-of-use electricity price is implemented with greater difference, energy stored system is suggested to be adopted for cooling and heating if off-peak electricity has the obvious effects of "peak load shifting" and operating cost conservation through technical and economical comparison.

　　11　In hot summer and cold winter zones, water-deficient areas, air source heat pump or soil source and ground source heat pump system is suggested to be adopted for cooling and heating in medium- and small-sized buildings.

　　12　Where there are such resources as natural surface water available or shallow underground water available and 100% recharging ensured, surface water or underground water ground source heat pump system may be adopted for cooling and heating.

　　13　In the areas with multiple energy resources, combined energy may be adopted for cooling and heating.

4.2.2　**The heating equipment powered by electric directly must not be adopted as heating source unless any one of the following conditions is met:**

　　1　**Power supply is sufficient and electricity utilization is encouraged by the management at the electricity demand side;**

　　2　**Buildings without urban or regional central heating where gas, coal, oil and other fuels are under restrictions of environmental protection or fire protection, and it is impossible to use heat pump as heating source;**

　　3　**The buildings primarily for cooling, with relatively very small heating loads, and incapable of using heat pump or other modes as heating source;**

　　4　**The air conditioning systems primarily for cooling, with small heating loads, and incapable of using heat pump or other modes as heating source, but using off-peak electricity for heat storage and electric boiler not a surge in electricity demand in peak hours and normal (flat) hours;**

　　5　**The buildings that can generate power by renewable energy and the power generation capacity can meet their own electric heating demands.**

4.2.3　**Direct electric heating equipment shall not be adopted as air humidification heating source unless any one of the following conditions is met:**

　　1　**Power supply is sufficient and electricity utilization is encouraged by the management at the electricity demand side;**

　　2　**The buildings that can generate power by renewable energy and the power generation capacity can meet their own humidifying power consumption demands.**

　　3　**The buildings without vapor source for humidifying in winter and with high accuracy control requirements for indoor relative humidity in winter.**

4.2.4　Boiler heating design shall meet the following requirements:

　　1　The design capacity of single set boiler shall be determined in the principle of ensuring its long-duration higher operating efficiency, the actual operating load rate is not suggested to be less than 50%;

　　2　Under the premise of ensuring that the boiler has long-duration higher operating efficiency, the capacities of all boilers are suggested to be equal;

　　3　Where the designed return water temperature of heating system is less than or equal to 50℃, condensing boilers are suggested to be adopted.

4.2.5　**Heat efficiency of boiler under nominal conditions and specified conditions must not be less than the values specified in Table 4.2.5.**

Table 4.2.5 Heat Efficiency of Boiler under Nominal Conditions and Specified Conditions (%)

Boiler type and fuel variety		Rated evaporating capacity of boiler D(t/h)/rated thermal power Q (MW)					
		$D<1/$ $Q<0.7$	$1\leqslant D\leqslant 2/$ $0.7\leqslant Q$ $\leqslant 1.4$	$2<D<6/$ $1.4<Q$ <4.2	$6\leqslant D\leqslant 8/$ $4.2\leqslant Q$ $\leqslant 5.6$	$8<D\leqslant 20/$ $5.6<Q$ $\leqslant 14.0$	$D>20/$ $Q>14.0$
Oil-fired and gas-fired boiler	Heavy oil	86	88				
	Light oil	88	90				
	Gas	88	90				
Layered burning boiler		75	78	80		81	82
Spreader-stoker-fired boiler	Category Ⅲ bituminous coal	—	—	—		82	83
Fluidized bed boiler		—	—	—		84	

4.2.6 Steam boiler shall not be adopted as the heating source except for the following conditions:

1 Steam thermal load must be adopted for kitchen, laundry, high-temperature sterilization and technological humidity control;

2 Steam thermal load is greater than 70% in the total thermal load that does not exceed 1.4MW.

4.2.7 Numbers of water chilling units (heat pumps) and a single set refrigerating (heating) capacity in central air conditioning system shall be selected according to the requirements of seasons and part loads and catering to the change law of loads all year around. Units are not suggested to be less than two sets and units of the same type should not exceed 4 sets; if only one set is arranged in small-scale engineering, the type with excellent regulation performance shall be selected and shall meet the requirements of minimum loads of buildings.

4.2.8 **Total installed capacity of motor-compression type chiller unit must be directly selected according to the air conditioning cooling load value calculated of requirement of article 4.1.1 in this standard and must not be added otherwise. Under designed conditions, if specification of the unit fails to meet the requirements of calculated cooling load, that ratio of the selected total installed capacity of the unit to the calculated cooling load must not be greater than 1.1.**

4.2.9 When distributed energy station is adopted as heating/cooling source, heat pump system driven by own power generation is suggested to be adopted that uses the waste heat from combined heat and power generation as low-grade heating source.

4.2.10 If motor-driven vapor compression cycle water chilling units (heat pumps) are adopted, **coefficient of performance (COP) under their nominal refrigeration and specified conditions must meet the following requirements:**

1 Coefficient of performance (COP) of water chilling unit at constant frequency and air-cooling or evaporative cooling unit shall not be less than the values specified in Table 4.2.10;

2 Coefficient of performance (COP) of centrifugal water chilling unit at variable frequency shall not be less than 0.93 times of the values specified in Table 4.2.10;

3 Coefficient of performance (COP) of water chilling screw unit at variable frequency shall not be less than 0.95 times of the values specified in Table 4.2.10.

Table 4.2.10 Coefficient of Performance (COP) of Water Chilling Units (Heat Pumps) under Their Nominal Refrigeration and Specified Conditions

Type		Nominal refrigerating capacity CC (kW)	Coefficient of performance COP(W/W)					
			Severe cold zones A and B	Severe cold zone C	Temperate zone	Cold zone	Hot summer and cold winter zone	Hot summer and warm winter zone
Water chilling	Piston/scroll	CC≤528	4.10	4.10	4.10	4.10	4.20	4.40
	Screw	CC≤528	4.60	4.70	4.70	4.70	4.80	4.90
		528<CC≤1163	5.00	5.00	5.00	5.10	5.20	5.30
		CC>1163	5.20	5.30	5.40	5.50	5.60	5.60
Water chilling	Centrifugal	CC≤1163	5.00	5.00	5.10	5.20	5.30	5.40
		1163<CC≤2110	5.30	5.40	5.40	5.50	5.60	5.70
		CC>2110	5.70	5.70	5.70	5.80	5.90	5.90
Air or evaporative cooling	Piston/scroll	CC≤50	2.60	2.60	2.60	2.60	2.70	2.80
		CC>50	2.80	2.80	2.80	2.80	2.90	2.90
	Screw	CC≤50	2.70	2.70	2.70	2.80	2.90	2.90
		CC>50	2.90	2.90	2.90	3.00	3.00	3.00

4.2.11 Integrated part load values (IPLV) of motor-driven vapor compression cycle water chilling units (heat pumps) shall meet the following requirements:

1 Calculation method of integrated part load value (IPLV) shall meet the requirements of article 4.2.13 in this standard;

2 Integrated part load values (IPLV) of water chilling unit at constant frequency shall not be less than the values in Table 4.2.11;

3 Integrated part load values (IPLV) of centrifugal water chilling unit at variable frequency shall not be less than 1.30 times of limits of centrifugal water chilling unit in Table 4.2.11;

4 Integrated part load values (IPLV) of water chilling screw unit at variable frequency shall not be less than 1.15 times of limits of water chilling screw unit in Table 4.2.11.

Table 4.2.11 Integrated Part Load Values (IPLV) of Water Chilling Units (Heat Pumps)

Type		Nominal refrigerating capacity CC(kW)	Integrated part load values IPLV					
			Severe cold zones A and B	Severe cold zone C	Temperate zone	Cold zone	Hot summer and cold winter zone	Hot summer and warm winter zone
Water chilling	Piston/scroll	CC≤528	4.90	4.90	4.90	4.90	5.05	5.25
Water chilling	Screw	CC≤528	5.35	5.45	5.45	5.45	5.55	5.65
		528<CC<1163	5.75	5.75	5.75	5.85	5.90	6.00
		CC>1163	5.85	5.95	6.10	6.20	6.30	6.30
	Centrifugal	CC≤1163	5.15	5.15	5.25	5.35	5.45	5.55
		1163<CC≤2110	5.40	5.50	5.55	5.60	5.75	5.85
		CC>2110	5.95	5.95	5.95	6.10	6.20	6.20

Table 4.2.11(continued)

Type		Nominal refrigerating capacity CC(kW)	Integrated part load values IPLV					
			Severe cold zones A and B	Severe cold zone C	Temperate zone	Cold zone	Hot summer and cold winter zone	Hot summer and warm winter zone
Air or evaporative cooling	Piston/scroll	CC≤50	3.10	3.10	3.10	3.10	3.20	3.20
		CC>50	3.35	3.35	3.35	3.35	3.40	3.45
	Screw	CC≤50	2.90	2.90	2.90	3.00	3.10	3.10
		CC>50	3.10	3.10	3.10	3.20	3.20	3.20

4.2.12 System coefficient of refrigeration performance (*SCOP*) of air conditioning system shall not be less than the values in Table 4.2.12. As for the chilled water systems composed of several water chilling units, cooling water pumps and cooling towers, comprehensive statistics and calculation shall be conducted for nominal refrigerating capacity and power consumption of all the equipment involved in operation. In the case of units in different types, their limits shall be determined according to weighting mode of the refrigerating capacity.

Table 4.2.12 System Coefficient of Refrigeration Performance (*SCOP*) of Air Conditioning System

Type		Nominal refrigerating capacity CC(kW)	System coefficient of refrigeration performance SCOP(W/W)					
			Severe cold zones A and B	Severe cold zone C	Temperate zone	Cold zone	Hot summer and cold winter zone	Hot summer and warm winter zone
Water chilling	Piston/scroll	CC≤528	3.3	3.3	3.3	3.3	3.4	3.6
	Screw	CC≤528	3.6	3.6	3.6	3.6	3.6	3.7
		528<CC<1163	4	4	4	4	4.1	4.1
		CC≥1163	4	4.1	4.2	4.4	4.4	4.4
	Centrifugal	CC≤1163	4	4	4	4.1	4.1	4.2
		1163<CC<2110	4.1	4.2	4.2	4.4	4.4	4.5
		CC≥2110	4.5	4.5	4.5	4.5	4.6	4.6

4.2.13 Integrated part load values (*IPLV*) of motor-driven vapor compression cycle water chilling units (heat pumps) shall be calculated according to the following formula:

$$IPLV = 1.2\% \times A + 32.8\% \times B + 39.7\% \times C + 26.3\% \times D \quad (4.2.13)$$

Where, A——the performance coefficient at 100% load (W/W), water intake temperature 30℃ of cooling water /inlet air dry bulb temperature 35℃ of condenser;

B——the performance coefficient at 75% load (W/W), water intake temperature 26℃ of cooling water /inlet air dry bulb temperature 31.5℃ of condenser;

C——the performance coefficient at 50% load (W/W), water intake temperature 23℃ of cooling water /inlet air dry bulb temperature 28℃ of condenser;

D——the performance coefficient at 25% load (W/W), waterintake temperature 19℃ of cooling water /inlet air dry bulb temperature 24.5℃ of condenser.

4.2.14 If motor-driven unitary air conditioner, duct air supply and rooftop air conditioning unit with nominal refrigerating capacity more than 7.1kW are adopted, energy efficiency ratio (*EER*) under nominal refrigeration and specified conditions must not be less than the values in Table 4.2.14.

Table 4.2.14 Energy Efficiency Ratio (*EER*) of Unitary Air Conditioner, Air Supply with Duct and Rooftop Air Conditioning Unit under Nominal Refrigeration and Specified Conditions

Type		Nominal refrigerating capacity CC(kW)	Energy efficiency ratio EER (W/W)					
			Severe cold zones A and B	Severe cold zone C	Temperate zone	Cold zone	Hot summer and cold winter zone	Hot summer and warm winter zone
Air cooling	Without air duct	$7.1<CC\leqslant14.0$	2.70	2.70	2.70	2.75	2.80	2.85
		$CC>14.0$	2.65	2.65	2.65	2.70	2.75	2.75
	With air duct	$7.1<CC\leqslant14.0$	2.50	2.50	2.50	2.55	2.60	2.60
		$CC>14.0$	2.45	2.45	2.45	2.50	2.55	2.55
Water chilling	Without air duct	$7.1<CC\leqslant14.0$	3.40	3.45	3.45	3.50	3.55	3.55
		$CC>14.0$	3.25	3.30	3.30	3.35	3.40	3.45
	With air duct	$7.1<CC\leqslant14.0$	3.10	3.10	3.15	3.20	3.25	3.25
		$CC>14.0$	3.00	3.00	3.05	3.10	3.15	3.20

4.2.15 Design of air source heat pump unit shall meet the following requirements:

1 With advanced and reliable defrosting control; total defrosting time shall not exceed 20% of the operation cycle;

2 Under the design conditions in winter, coefficient of performance (*COP*) of cold/hot air unit shall not be less than 1.8; coefficient of performance (*COP*) of water chiller-heater unit shall not be less than 2.0;

3 Auxiliary heating sources shall be arranged, when outdoor designed temperature is lower than local temperature at balance point or room temperature stability has relatively high requirements in the cold and damp area in winter;

4 Heat-recovery heat pump unit is suggested to be adopted for the buildings needto be cooled and heated simultaneously.

4.2.16 Arrangement of outdoor units of air source, air-cooled, evaporative-cooling water chilling units (heat pumps) shall meet the following requirements:

1 Ensuring smooth air intake and exhaust, no obvious air flow short circuit between air exhaust and air intake;

2 Prevention of impact of polluted air flow;

3 Noise and heat exhaust shall meet the requirements of the surrounding environment;

4 Shall be convenient for cleaning heat exchangerof outdoor unit.

4.2.17 When multi-connected air-conditioning (heat pump) unit is adopted, its cooling integrated part load value [*IPLV*(C)] under nominal refrigeration and specified conditions must not be less than the values in Table 4.2.17.

Table 4.2.17 Cooling Integrated Part Load Value [IPLV(C)] of Multi-connected Air-conditioning (Heat Pump) Unit under Nominal Refrigeration and Specified Conditions

Nominal refrigerating capacity CC(kW)	Cooling integrated part load value, IPLV(C)					
	Severe cold zones A and B	Severe cold zone C	Temperate zone	Cold zone	Hot summer and cold winter zone	Hot summer and warm winter zone
CC≤28	3.80	3.85	3.85	3.90	4.00	4.00
28<CC≤84	3.75	3.80	3.80	3.85	3.95	3.95
CC>84	3.65	3.70	3.70	3.75	3.80	3.80

4.2.18 Except for the VRV systems with heat recovery functional type or low-temperature heat pump type, equivalent length of refrigerant connecting pipe of VRV air conditioning system shall meet the requirement for energy efficiency ratio (EER) no less than 2.8 at full load corresponding to cooling conditions.

4.2.19 If direct-fired lithium bromide absorption chiller-heater unit is adopted, performance parameters under their nominal conditions and specified conditions must be in accordance with those specified in Table 4.2.19.

Table 4.2.19 Performance Parameters of Direct-fired Lithium Bromide Absorption Chiller-heater unit under Their Nominal Conditions and Specified Conditions

Nominal conditions		Performance parameters	
Inlet/outlet temperature for chilled (warm) water(℃)	Inlet/outlet temperature for cooling water (℃)	Coefficient of performance(W/W)	
		Cooling	Heating
12/7 (Cooling)	30/35	≥1.20	—
—/60 (Heating)	—	—	≥0.90

4.2.20 In the buildings with cooling demands in winter or transition season, fresh air shall be taken full advantage of for cooling; if the technical and economical analysis is reasonable, it is possible to use cooling tower to provide chilled water for air conditioning or air conditioner (heat pump) with simultaneous refrigeration and heating functions.

4.2.21 When steam is adopted as heating source, and the technical and economical comparison is reasonable, condensed water generated by steam-using equipment shall be recovered. Closed system shall be adopted for condensate recovery system.

4.2.22 For the buildings with domestic hot water demand all year around, the water chilling unit with condensation heat recovering function is suggested to be adopted if motor vapor compression cycle water chilling unit is adopted.

4.3 Transmission and Distribution System

4.3.1 Hot water shall be adopted as heating medium for central heating system.

4.3.2 Hydraulic balancing device shall be set at heating entry of central heating system and branch of water supply pipe or water return pipe in accordance with the requirements of hydraulic balance.

4.3.3 In matching circulating water pump for central heating system, the electricity consumption to transferred heat quantity ratio (EHR-h) of central heating system shall be calculated, and it shall be indicated in the design specification of construction drawing. Electricity consumption to transferred heat quantity ratio of central heating system shall be calculated according to the following formula:

$$EHR\text{-}h = 0.003096 \Sigma(G \times H/\eta_b)/Q \leqslant A(B + \alpha \Sigma L)/\Delta T \qquad (4.3.3)$$

Where, $EHR\text{-}h$——the electricity consumption to transferred heat quantity ratio of central heating system;

G——the designed flow of each pump in operation (m³/h);

H——the designed water-head of each pump in operation (mH$_2$O);

η_b——the designed working point efficiency of each pump in operation;

Q——the designed thermal load (kW);

ΔT——the designed temperature difference between supply water and return water (℃);

A——the calculation coefficient related to pump flow, which is selected from Table 4.3.9-2 of this standard;

B——the calculation coefficient related to the water resistance of equipment room and the user, B as 17 taken for primary pump system, B as 21 for secondary pump system;

ΣL——the total length of supply and return water pipes from heating substation to heating terminal (radiator or water collector and distributor for radiant heating) (m);

α——the calculation coefficient related to ΣL;

Where $\Sigma L \leqslant 400$m, $\alpha = 0.0115$;

Where $400\text{m} < \Sigma L < 1000\text{m}$, $\alpha = 0.003833 + 3.067/\Sigma L$;

Where $\Sigma L \geqslant 1000$m, $\alpha = 0.0069$.

4.3.4 Where variable flow water system is adopted for the central heating system, circulating water pump is suggested to be controlled at variable speed adjustment.

4.3.5 Chilled and hot water systems for central air conditioning shall be designed in accordance with the following requirements:

1 Where seasonal transfer of simultaneous cooling and heating is only required in all zones of a building, two-pipe air conditioning water system shall be adopted; where full-year cooling is required for air conditioning system in some zones of a building and seasonal transfer of cooling and heating is only required in other zones, zoned two-pipe air conditioning water system may be adopted; where cooling and heating conditions of air conditioning water system are frequently transferred or require to be used simultaneously, four-pipe air conditioning water system is suggested to be adopted.

2 When chilled watertemperature and temperature difference of supply/return water is required to be consistent, and for small and medium-sized works in each zone without much different pipe pressure loss, variable-flow primary pump system is suggested to be adopted; where a single water pump has larger power, through comparison of technology and economy, variable flow primary pump system may be adopted at water chilling unit and loading side for air conditioning chilled water under the premise of ensuring the adaptability of equipment and

reliability of control scheme and operation management, and speed governing pump shall be adopted for primary pump.

3 For major works with the working radius of the system being larger and the designed flow resistance being higher, variable flow secondary pump system is suggested to be adopted for air conditioning chilled water. Where designed water temperature of each loop is consistent and the designed flow resistance is approximated, the secondary pumps should be arranged in a centralized manner; where the designed flow resistance of each loop is much different or water temperature or difference of each system is differently required, the secondary pumps should be arranged by zone or system respectively, and variable speed pump shall be adopted for secondary pump;

4 For the large-scale air conditioning chilled water system with cooling source equipment centrally arranged and user-decentralized district cooling, where conveying distance of the secondary pump is far away and the difference between the users' pipe resistances is larger, or the requirement for water temperature (temperature difference) is different, multi-stage pump system may be adopted, and variable speed pump shall be adopted for pumps at all levels at the loading side of the secondary pump.

4.3.6 Arrangement of air conditioning water system and selection of pipe diameter shall lead to the reduction of relative difference in pressure loss between parallel loops. Where the relative difference in pressure loss between parallel loops under designed conditions exceeds 15%, hydraulic balance measures shall be taken.

4.3.7 Circulating water pump heated by heat exchanger or that of cooled secondary air conditioning water system is suggested to be controlled at variable speed.

4.3.8 Besides that designed flow, resistance characteristics of pipe network and working characteristics of pump for air conditioning chilled water system and hot water system are close, the chilled water and hot water circulating pumps shall be arranged respectively for the two-pipe air conditioning water system.

4.3.9 In matching circulating water pump for air conditioning chilled and hot water system, the electricity consumption to transferred cooling (heating) quantity ratio $[EC(H)R\text{-}a]$ of air conditioning chilled/hot water system shall be calculated and marked in the design description of construction drawing. The electricity consumption to transferred cooling (heating) quantity ratio of air conditioning chilled/hot water system shall meet the following requirements:

1 The electricity consumption to transferred cooling (heating) quantity ratio of air conditioning chilled/hot water system shall be calculated according to the following formula:

$$EC(H)R\text{-}a = 0.003096 \sum(G \times H/\eta_b)/Q \leqslant A(B + \alpha \sum L)/\Delta T \qquad (4.3.9)$$

Where, $EC(H)R\text{-}a$——the ratio of electricity consumption to transferred cooling (heating) quantity of air conditioning chilled/hot water system circulating water pump;

G——the designed flow of each pump in operation (m³/h);

H——the water head of each pump in operation (mH₂O);

η_b——the designed working point efficiency of each pump in operation;

Q——the designed cooling (heating) load (kW);

ΔT——the specified supply and return water temperature difference for calculation, which is selected from Table 4.3.9-1;

A —— the calculation coefficient related to pump flow, which is selected from Table 4.3.9-2;

B —— the calculation coefficient related to the equipment room and user's water resistance, which is selected from Table 4.3.9-3;

α —— the calculation coefficient related to ΣL, which is selected from Table 4.3.9-4 or 4.3.9-5;

ΣL —— the total conveying length of supply and return water pipe from the exit of quipment room of cooling/heating source to the furthest user of this system (m).

Table 4.3.9-1 ΔT Values (℃)

Chilled water system	Hot water system			
	Severe cold	Cold	Hot summer and cold winter	Hot summer and warm winter
5	15	15	10	5

Table 4.3.9-2 A Values

Designed pump flow G	$G \leqslant 60 m^3/h$	$60 m^3/h < G \leqslant 200 m^3/h$	$G > 200 m^3/h$
A value	0.004225	0.003858	0.003749

Table 4.3.9-3 B Values

System composition		B value for four-pipe single-cold, single-hot piping	B value for two-pipe hot water piping
One stage pump	Chilled water system	28	—
	Hot water system	22	21
Two stage pump	Chilled water system	33	—
	Hot water system	27	25

Table 4.3.9-4 α Values for Four-pipe Chilled and Hot Water Piping System

System	Pipe length (ΣL) range (m)		
	$\Sigma L \leqslant 400 m$	$400 m < \Sigma L < 1000 m$	$\Sigma L \geqslant 1000 m$
Chilled water system	$\alpha = 0.02$	$\alpha = 0.016 + 1.6/\Sigma L$	$\alpha = 0.013 + 4.6/\Sigma L$
Hot water system	$\alpha = 0.014$	$\alpha = 0.0125 + 0.6/\Sigma L$	$\alpha = 0.009 + 4.1/\Sigma L$

Table 4.3.9-5 α Values for Two-pipe Hot Water Pipe System

System	zone	Pipe length (ΣL) range (m)		
		$\Sigma L \leqslant 400 m$	$400 m < \Sigma L < 1000 m$	$\Sigma L \geqslant 1000 m$
Hot water system	Severe cold	$\alpha = 0.009$	$\alpha = 0.0072 + 0.72/\Sigma L$	$\alpha = 0.0059 + 2.02/\Sigma L$
	Cold	$\alpha = 0.0024$	$\alpha = 0.002 + 0.16/\Sigma L$	$\alpha = 0.0016 + 0.56/\Sigma L$
	Hot summer and cold winter	$\alpha = 0.0032$	$\alpha = 0.0026 + 0.24/\Sigma L$	$\alpha = 0.0021 + 0.74/\Sigma L$
	Hot summer and warm winter			
Chilled water system		$\alpha = 0.02$	$\alpha = 0.016 + 1.6/\Sigma L$	$\alpha = 0.013 + 4.6/\Sigma L$

2 Calculation parameter for the electricity consumption to transferred cooling (heating) quantity ratio of air conditioning chilled/hot water system shall meet the following requirements:

  1) Temperature difference of hot supply and return water for air-source heat pump, lithium bromide unit and water source heat pump shall be determined by actual parameters of the units; that of chilled supply and return water for units directly supplying high-temperature chilled water shall be determined by the actual parameters of the units.
  2) Where multi-set pumps operate in parallel, the A value shall be selected by the larger flow.
  3) The B value for two-pipe chilled water piping shall be selected by that value for four-pipe single-cold piping; for chilled water system with multistage pumps, for each additional stage of pump, the B value may be increased by 5; for hot water system with multistage pump, for each additional stage of pump, the B value may be increased by 4.
  4) Formula for calculation of α for two-pipe chilled water system shall be the same as that for four-pipe chilled water system.
  5) If the furthest user is to be fan coil, ΣL shall be determined by the total length of supply and return water pipe from the exit of equipment room to the farthest fan coil minus 100m.

4.3.10 If a long-time use ventilation system's operating condition (air volume and wind pressure) has great changes, the double speed fan or variable speed fan is suggested to be adopted for ventilation.

4.3.11 In the design period of all-air conditioning system with constant air volume, measures to realize the operation of all fresh air or adjust fresh air ratio is suggested to be taken, and corresponding ventilation system is suggested to be designed.

4.3.12 If an air conditioning wind system works for several service spaces, its fresh air volume shall be calculated according to the following formulae:

$$Y = X/(1+X-Z) \quad (4.3.12\text{-}1)$$
$$Y = V_{ot}/V_{st} \quad (4.3.12\text{-}2)$$
$$X = V_{on}/V_{st} \quad (4.3.12\text{-}3)$$
$$Z = V_{oc}/V_{sc} \quad (4.3.12\text{-}4)$$

Where, Y——the corrected ratio of system fresh air rate to air supply rate;
  V_{ot}——the corrected total fresh air rate (m³/h);
  V_{st}——the total air supply rate, namely the sum of air supply rate of the system in all rooms (m³/h);
  X——the uncorrected ratio of system fresh air rate to air supply rate;
  V_{on}——the sum of fresh air rate of the system in all rooms (m³/h);
  Z——the fresh air ratio in the room with the greatest demand for fresh air ratio;
  V_{oc}——the fresh air rate in the room with the greatest demand for fresh air ratio (m³/h);
  V_{sc}——the air supply rate in the room with the greatest demand for fresh air ratio (m³/h).

4.3.13 For rooms with relatively large and variable population density, fresh air demand is suggested to be controlled according to the detection value of indoor CO_2 concentration and exhaust air volume is suggested to adapt to the variation of fresh air rate so as to keep positive pressure in the rooms.

4.3.14 If artificial cooling and heating sources are adopted to preheat or precool air conditioning system, fresh air system shall be able to be closed down; if the outdoor air temperature is low fairly, fresh air system shall be used as possible for precooling.

4.3.15 Internal and external zones for air conditioning shall be divided by indoor depth, compartment, orientation, storey and building envelope characteristics. Air conditioning system should be arranged respectively in these zones.

4.3.16 Fresh air for primary air fan-coil systems is suggested to be directly supplied to each conditioned zone and not to be exhausted through fan coil unit.

4.3.17 Air filter shall be designed in accordance with the following requirements:

   1 Performance parameters of air filter shall meet the requirements of the current national standard of GB/T 14295 *Air Filter*;

   2 Monitoring and alarming devices for filter resistance are suggested to be arranged and shall be of replacement conditions;

   3 Filter for full-airair conditioning system shall be able to meet the demand for operation of all fresh air.

4.3.18 Air conditioning wind systems shall not use civil engineering flue as supply air duct and convey duct of conditioned fresh air. If civil engineering flue is used due to limited conditions, reliable measure for air leakage prevention and heat insulation shall be taken.

4.3.19 Air conditioning cooling water systems shall be designed in accordance with the following requirements:

   1 It shall have functions of water treatment, such as filtration, corrosion mitigation, scale inhibition, sterilization and algae killing.

   2 Cooling towers are suggesed to beinstalled in the place with good air flow conditions;

   3 Water flowmeter shall be set on water supplement main of cooling tower;

   4 When cooling water collection tank is arranged indoors, the difference of design water level between water distributor and water collection tank of the cooling tower shall not be greater than 8m.

4.3.20 Supply air temperature difference of air conditioning system shall be determined through calculation according to air handling procedure expressed in enthalpy-humidity diagram. If upper air supply air flow is adopted for air conditioning system, the design supply air temperature difference is suggested to be increased in summer and shall meet the following requirements:

   1 Where the height of air supply is less than or equal to 5m, supply air temperature difference is not suggested to be less than 5℃;

   2 Where the height of air supply is greater than 5m, supply air temperature difference is not suggested to be less than 10℃.

4.3.21 Heating and cooling process is not suggested to be carried out simultaneously in the same air handling system.

4.3.22 Where air volume of air conditioning wind system and ventilation system is greater than 10,000m³/h, energy consumption per unit air volume of duct system (W_s) is not suggested to be greater than the value specified in Table 4.3.22. The energy consumption per unit air volume of duct system (W_s) shall be calculated according to the following formula:

$$W_s = P/(3600 \times \eta_{CD} \times \eta_F) \qquad (4.3.22)$$

Where, W_s——the energy consumption per unit air volume of duct system [W/(m³/h)];
  P——the residual pressure of air conditioning unit or the wind pressure of ventilation system's fan (Pa);
  η_{CD}——the electric motor and transmission efficiency (%), 0.855 is taken;
  η_F——the fan efficiency (%), which shall be selected according the efficiency indicated in design drawing.

Table 4.3.22 Energy Consumption per Unit Air Volume of Duct System W_s[W/(m³/h)]

System form	Limit of W_s
Mechanical ventilation system	0.27
Fresh air system	0.24
Constant air volume system of office buildings	0.27
Variable air volume system of office buildings	0.29
All air system of commercial and hotel buildings	0.30

4.3.23 Where cooling medium delivery temperature is less than the environmental temperature outside the pipe and cooling medium temperature is not allowed to rise, or where heating medium delivery temperature is higher than the environmental temperature outside the pipe and heating medium temperature is not allowed to drop, thermal insulation measures shall be taken for pipes and equipment. insulation layer shall be set in accordance with the following requirements:

  1 The thickness of the insulation layer shall be calculated by the method of economic thickness specified in the current national standard of GB/T 8175 *Guide for Design of Thermal Insulation of Equipments and Pipes*;

  2 Where cooling or cooling and heating is carried out at the same time, the thickness of the insulation layer shall be calculated by economic thickness and the thickness of cold insulation layer avoiding surface dew formation in the current national standard of GB/T 8175 *Guide for Design of Thermal Insulation of Equipments and Pipes* and the larger value shall be taken;

  3 Insulation layer thickness for pipe and equipment and minimum thermal resistance of air duct insulation layer may be selected according to the requirements of Appendix D in this standard;

  4 "thermal bridge" or "cold bridge" prevention measures shall be taken between the pipes and supports, in the pipe through-wall and through-floor slab;

  5 If non-obturator material is adopted for thermal insulation, protection layer shall be set on external surface; if it is adopted for cold insulation, vapor barrier coating and protection layer shall be set on the external surface.

4.3.24 Electric air valves shall be set on the duct and device of ventilation system or air conditioning system in severe cold and cold zones, which shall have the function of automatic chain closing and tight sealing, and sealing measures shall be taken.

4.3.25 If the air conditioning system with assemble exhaust is deemed as reasonable through technical and economical comparison, an air-air energy recovery device is suggested to be set. If the device is adopted in severe cold zone, the check shall be taken whether its air exhaust side is subject to frosting or dewing. In case of frosting or dewing, preheating and other insulation antifreeze measures should be taken.

4.3.26 With the function of heat recovery two-way ventilation devices is suggested to be installed respectively in each conditioned zone or air-conditioned room where personnel have a long stay without centralized air ventilation systems.

4.4 Terminal System

4.4.1 Radiator is suggested to be installed in an exposed way; thermal resistance of material for radiant heating surface of floor is not suggested to be greater than $0.05 m^2 \cdot K/W$.

4.4.2 In areas with low outdoor calculated wet-bulb temperature for summer air conditioning and large daily range of temperature, direct evaporative cooling, indirect evaporative cooling or both combined secondary or tertiary evaporative cooling is suggested to be adopted for air handling.

4.4.3 In designing VAV full-air conditioning system, the way of automatically adjusting the speed of fan at variable frequency shall be adopted and the minimum air supply volume of each VAV terminal device shall be indicated in design document.

4.4.4 Where the height of building space is greater than or equal to 10m and its volume is greater than $10,000 m^3$, radiant heating and cooling system or stratified air conditioning system is suggested to be adopted.

4.4.5 Ventilation design of electromechanical equipment room and kitchen hot processing room with larger heat productivity shall meet the following requirements:

1 Under the premise of ensuring normal operation of equipment, rooms are suggested to be ventilated to eliminate residual heat. Indoor design temperature in summer of electromechanical equipment room is not suggested to be lower than the outdoor calculated temperature for summer ventilation.

2 Air supplement type exhaust hood is suggested to be adopted in kitchen hot processing room. In direct airflow ventilated zones, indoor calculated temperature in summer is not suggested to be lower than the outdoor calculated temperature for summer ventilation.

4.5 Monitor, Control and Measure

4.5.1 Central heating ventilation and air conditioning system (HVAC) shall be monitored and controlled. Where all air conditioning system is used in public buildings with area greater than $20,000 m^2$, direct digital control system is suggested to be adopted. System function and monitoring control contents shall be determined through technical and economical comparison according to building function, relevant standard and system type.

4.5.2 **Energy metering must be carried out for boiler room, heat exchanger room and refrigerator room, which must include the following:**

1 **Fuel consumption;**

2 **Power consumption of refrigerator;**

3 **Heat supply of central heating system;**

4 **Water supplement.**

4.5.3 Where regional cooling source and heating source are adopted, cooling and heat metering device shall be set at the entrance of cooling and heating sources of every public building. Where central heating and air conditioning system is adopted, cooling metering device and heat metering device are suggested to be set respectively in different user units or zones.

4.5.4 **Automatic heat supply controller must be set in boiler room and heat exchanger room.**

4.5.5 Control design of boiler room and heat exchanger room shall meet the following requirements:

  1 Pump shall be interlocking controlled with valve etc. equipment;

  2 Supply water temperature shall be regulated according to outdoor temperature;

  3 Water supply flowrate shall be regulated according to the demand of terminal;

  4 The number of pumps and rotating speed is suggested to be controlled according to the demand of terminal;

  5 The number of boilers put into operation and the fuel consumption shall be controlled according to the demand for heat supply.

4.5.6 Indoor temperature control device must be set in heating and air conditioning system; automatic temperature control valve shall be installed in radiator and radiant heating system.

4.5.7 Control function of cooling/heating source equipment room shall meet the following requirements:

  1 Cold water (heat pump) unit, water pump, valve, cooling tower etc. equipment shall have the function of procedure start-up/shut-down and coordinated control;

  2 The unit number control for water chillers shall be available and optimal control of cooling capacity is suggested to be adopted;

  3 The unit number control for water pumps shall be available and optimal control of flowrate is suggested to be adopted;

  4 The secondary pump shall be able to be controlled through automatic variable speed and the speed of rotation is suggested to be controlled according to the differential pressure of pipe, and the differential pressure is suggested to be optimized;

  5 The unit number control for cooling tower fans shall be available and variable speed control is suggested to be carried out according to outdoor meteorological parameters;

  6 Automatic sewerage discharge of cooling tower shall be able to be controlled;

  7 Supply water temperature is suggested to be optimized according to the outdoor meteorological parameters and the demand of terminal;

  8 Equipment is suggested to be able to be rotated for use according to the accumulated runtime;

  9 If there are more than 3 sets cooling/heating source main machines, group control of unit is suggested to be adopted; if group control is adopted, control system shall establish communication link with control unit contained in water chilling unit.

4.5.8 All air conditioning system shall be controlled in accordance with the following requirements:

  1 Fan, air valve and water valve shall be able to be started/stopped through interlocking control;

  2 It shall be periodically started/stopped according to service time, and the start/stop time should be optimized;

  3 Where variable air volume system is adopted, the fan shall be controlled at variable speed;

  4 The control mode of increasing fresh air ratio is suggested to be adopted in transition season;

  5 Set value of indoor temperature is suggested to be optimized according to outdoor meteorological parameters;

  6 Control mode of delayed closing in the absence of personnel is suggested to be set at air supply terminal of all fresh air system.

4.5.9 Fan coil shall be controlled by both electric water valve and air speed, and normally-closed type electric on-off valve is suggested to be set. Fan coil in public area shall be controlled in accordance with the following requirements:

  1 Range of set value of indoor temperature shall be able to be limited;

  2 It shall be periodically started/stopped according to service time, and the start/stop time is suggested to be optimized.

4.5.10 For ventilation system mainly aiming to eliminate indoor waste heat, the number of operating ventilation equipment or its rotation speed is suggested to be controlled according to room temperature.

4.5.11 Fans in underground garages is suggested to be connected in multi-set parallel or speed regulator for them is suggested to be set, and the fans are suggested to be periodically started/stopped (the number of fans) according to service condition or automatic operation control is suggested to be carried out according to the carbon monoxide concentration in the garage.

4.5.12 Automatic start/stop controller is suggested to be set in air conditioning system in intermittent operation. The controller shall be capable of starting/stopping the equipment according to preset schedule and the presence of person in service area.

5 Water Supply and Drainage

5.1 General Requirements

5.1.1 Water saving design of water supply and drainage system shall meet the requirements of the current national standards of GB 50015 *Code for Design of Building Water Supply and Drainage* and GB 50555 *Standard for Water Saving Design in Civil Building*.

5.1.2 Water meter for measurement shall be set according to building type, water using department and management requirements, and shall meet the requirements of the current national standard of GB 50555 *Standard for Water Saving Design in Civil Building*.

5.1.3 Hot water meter, heat meter, vapor flow meter or energy meter shall be installed in water heating room and heat exchange station room requiring meter.

5.1.4 The type of feed pump shall be selected according to the hydraulic computation result of supply pipe network and shall ensure that water pump efficiency lies in high efficiency area under designed conditions. The efficiency of feed pump is not suggested to be less than the evaluating values of energy conservation for pumps as specified in the current national standard of GB 19762 *Limited Values of Energy Efficiency and Evaluating Values of Energy Conservation of Centrifugal Pump for Fresh Water*.

5.1.5 Sanitary wares and accessories in toilet shall meet the requirements of the current national standard of CJ/T 164 *Domestic Water Saving Devices*.

5.2 Water Supply and Drainage System

5.2.1 Water supply system shall supply water directly through water pressure of town water supply network or community water supply network. Overlying water supply system may be adopted upon approval.

5.2.2 The number, scale, location of secondary booster pump stations and the water supply pressure of pump unit shall be reasonably determined according to town water supply conditions, community scale, building height, building distribution, service standard, safe water supply and energy consumption reduction.

5.2.3 Water supply mode and vertical zoning of water supply system shall be determined according to use, number of storeys, service requirements, performance of materials and equipment, maintenance management and energy consumption of buildings. Pressure requirements of zones shall meet the requirements of the current national standards of GB 50015 *Code for Design of Building Water Supply and Drainage* and GB 50555 *Standard for Water Saving Design in Civil Building*.

5.2.4 Variable frequency-speed pump unit shall be reasonably selected according to water consumption and water consumption uniformity so as to match water pump and regulating facilities, and the started number of water pumps is suggested to be automatically controlled according to the demand for water supply so as to ensure operation in high efficiency area.

5.2.5 Domestic sewage and waste water over ground is suggested to be discharged into outdoor pipe network directly through gravity flow system.

5.3 Service Water Heating

5.3.1 Residual heat, waste heat, renewable energy or air source heat pump are suggested to be used as heating source of central hot water supply system. If the maximum daily domestic hot water yield is greater than 5m^3, direct electrical heating source shall not be adopted as the heating source of central hot water supply system, except for the electricity utilization encouraged by the management at the electricity demand side and that of valley power for heating.

5.3.2 Where gas or fuel oil is used as heating source, gas or fuel oil unit is suggested to be adopted to directly prepare hot water. Where the boiler is used to prepare domestic hot water or boiled water, thermal efficiency of the boiler under rated conditions shall not be less than the limited values specified in Table 4.2.5 of this standard.

5.3.3 Where air-source heat pump water heater unit is adopted to prepare domestic hot water, coefficient of performance (COP) of heat pump water heater with heating capacity greater than 10kW under nominal heating condition and specified conditions should not be less than the requirements of Table 5.3.3, and effective measures shall be taken to ensure water quality.

Table 5.3.3 Coefficient of Performance (COP) of Heat Pump Water Heater (W/W)

Heating capacity H(kW)	Type of water heater		Ordinary type	Low temperature type
H≥10	Direct-heating type		4.40	3.70
	Cyclic heating	Without water pump	4.40	3.70
		With water pump	4.30	3.60

5.3.4 Service radius of hot water circulating pipe network of central hot water supply system in the community is neither suggested to be greater than 300m or be greater than 500m. Water heating room and heat exchange station room is suggested to be set at the center of the community.

5.3.5 Centralized domestic hot water supply system should not be set in the building only equipped with hand basin. Separate hot water circulation system is suggested to be set in the building equipped with central hot water supply system, whose designed value of daily hot water consumption is greater than or equal to 5m^3, or to set for uses constantly supplied with hot water.

5.3.6 Water supply zone of central hot water supply system is suggested to be the same as chilled water zone at water consuming point, and relevant measures shall be taken to ensure the balance of chilled and hot water supply pressure at the water consuming point and the effective circulation of circulation pipe network.

5.3.7 Insulation measures shall be taken for pipe network and equipment of central hot water supply system, and the thickness of insulation layer shall be determined by the calculation method for economic thickness specified in GB/T 8175 *Guide for Design of Thermal Insulation of Equipments and Pipes*, and it may also be selected according to the requirements of Appendix D in this standard.

5.3.8 Central hot water supply system is suggested to be monitored and controlled according to the following requirements:

1 Hot water consumption and total supply heat of the system is suggested to be monitored;

2 Equipment operation state is suggested to be tested and alarmed for malfunction;

3 Daily water consumption and supply water temperature is suggested to be monitored;

4 The project with installation quantity greater than or equal to 3 sets is suggested to be controlled in machine group.

6 Electric

6.1 General Requirements

6.1.1 Electrical system shall be designed in an economical rationality and high efficiency and energy-efficient manner.

6.1.2 Technology-advanced, sophisticated, reliable economical and energy-efficient products with small amount of harmonic emission measure are suggested to be selected for electrical system.

6.1.3 Building automation system shall be set in accordance with the relevant requirements of the current national standard of GB 50314 *Standard for Design of Intelligent Building*.

6.2 Power Supply and Distribution System

6.2.1 Electrical system shall be designed according to local power supply conditions and supply voltage grade shall be reasonably determined.

6.2.2 Distribution substation shall be close to load center and high-power electric equipments.

6.2.3 Low-loss type shall be selected for transformer, and its energy efficiency value shall not be less than evaluating values of energy conservation in energy efficiency standard in the current national standard of GB 20052 *Minimum Allowable Values of Energy Efficiency and Energy Efficiency Grades for Three-phase Distribution Transformers*.

6.2.4 Transformers should be designed to ensure its operating within the range of economical operation parameters.

6.2.5 Unbalancedness of three-phase load for power distribution system is not suggested to be greater than 15%. For power supply system with more single load, Part split-phase reactive automatic compensation devices should be adopted.

6.2.6 For electric equipment with larger capacity, if power factor is lower and the equipment is far away from power substation and distribution substation, local reactive power compensation device is suggested to be adopted.

6.2.7 For equipment with large harmonic source, such as large electric equipment, silicon controlled dimmer, motor variable frequency speed control device, local harmonic suppression device is suggested to be set. If there is more nonlinear electric equipment in buildings, space is suggested to be reserved for installation of wave filtering.

6.3 Lighting

6.3.1 The indoor lighting power density (*LPD*) shall meet the requirements of the current national standard of GB 50034 *Standard for Lighting Design of Buildings*.

6.3.2 The energy efficiency of light source and ballast for design selection is not suggested to be less than the energy saving evaluation in corresponding energy efficiency standard.

6.3.3 The limit of lighting power density (*LPD*) for building nightscape shall meet the requirements of the current professional standard of JGJ/T 163 *Code for Lighting Design of Urban Nightscape*.

6.3.4 Light source shall be selected in accordance with the following requirements:

1 Light source with larger single-lamp power and higher lighting effect is suggested to be adopted for general lighting under meeting the condition of illumination uniformity, fluorescent high-pressure mercury lamp is not suggested to be selected and self-ballasted fluorescent high-pressure mercury lamp shall not be selected.

2 The ballast with low harmonic content shall be selected for gas discharge lamp;

3 Metal halide lamp and high-pressure sodium lamp is suggested to be selected in high and large space and outdoor work site;

4 Incandescent lamp shall not be adopted, except for the sites required special process requirements;

5 Light emitting diode (LED) lamp is suggested to be adopted in aisle, staircase, toilet, garage and other places in which men stay for a while;

6 Light emitting diode (LED) lamp is suggested to be selected for evacuation strobe light, exit sign lamp and indoor directive decorative lighting;

7 Safe, high-efficient, long-life and stable light source is suggested to be selected for outdoor landscape and road lighting so as to avoid light pollution.

6.3.5 Lamp fixture shall be selected in accordance with the following requirements:

1 For gas discharge lamp with inductive ballast, single-lamp compensation mode shall be adopted and power factor of its lighting distribution system shall not be less than 0.9;

2 High efficient lamp fixture shall be selected in case glare limitation and photometric performance requirements are met and shall meet the requirements of the current national standard of GB 50034 *Standard for Lighting Design of Buildings*;

3 Interface with lighting control system is suggested to be reserved for single lamp control device containing lamp fixture.

6.3.6 Mixed lighting is suggested to be adopted for such sites where general lighting fails to meet the requirement of luminance.

6.3.7 Diffused luminous ceiling is not suggested to be adopted for lighting design.

6.3.8 Lighting control shall meet the following requirements:

1 Lighting control shall be performed by zoning and grouping in combination with building service condition and natural daylight condition;

2 Energy saving type master switch shall be set in hotel room;

3 Except for rooms with single lamp fixture, at least 2 lamps fixture is suggested to be set in each room and the number of light sources each switch controls is not suggested to be greater than 6;

4 Centralized switch control or local inductive control is suggested to be adopted for lighting of corridor, staircase, hallway, elevator lobby, toilet, parking garage and other public places;

5 Intelligent lighting control system is suggested to be adopted for lighting of large space and multi-functional place with several scenarios;

6 In case of arranging electric shading device, luminance control is suggested to be interlocked with it;

7 Multiple-mode (at ordinary times, in general festivals and great festivals) automatic control device shall be set for landscape lighting of buildings.

6.4 Electric Power Supervision and Measure

6.4.1 Where electricity consumption of main secondary energy consumption organization is greater than or equal to 10kW or that of single set electrical equipment is greater than or equal to 100kW, electric energy meter shall be set. Power consumption monitoring and measurement system is suggested to be set in public buildings and energy efficiency shall be analyzed and managed.

6.4.2 Energy monitor and measurement system shall be set in public buildings according to functional area.

6.4.3 Electric power shall be monitored and measured in public buildings according to lighting socket, air conditioner, electricity and special electrical subitems. For office buildings, electric power is suggested to be monitored and measured according to lighting subitems and socket sub items.

6.4.4 Power consumption of circulating water pump for cooling/heating source system is suggested to be measured separately.

7 Renewable Energy Application

7.1 General Requirements

7.1.1 Renewable energy shall be preferentially used in public buildings through the analysis of local environment and resource conditions and technical and economic analysis in combination with relevant national policies.

7.1.2 Facilities for utilization of renewable energy for public buildings shall be synchronously designed with main building of a project.

7.1.3 When permissible environmental conditions allow and technical economy is rational, solar energy, wind energy and other renewable energy is suggested to be adopted for direct grid connection of power supply.

7.1.4 When public power grid fails to provide lighting source, solar energy and wind energy shall be adopted for power generation and storage battery shall be configured as lighting source.

7.1.5 Energy efficiency meter is suggested to be set for renewable energy application system.

7.2 Solar Energy Application

7.2.1 Solar energy shall be applied according to the principle of passive priority. It is suggested to be fully used for public building design.

7.2.2 Photothermic or photovoltaic with building integrated system should be adopted for public buildings; Photothermic or photovoltaic with building integrated system shall not affect the function of building external envelope structure and shall meet relevant requirements of current standards of the nation.

7.2.3 Solar photovoltaic solar-thermal integration system is suggested to be adopted when solar energy is utilized to supply heat and power for public buildings.

7.2.4 Solar fraction shall meet the requirements of Table 7.2.4 when solar thermal utilization system is set for public buildings.

Table 7.2.4 Solar Fraction f (%)

Zoning of solar energy resources	Solar water heating system	Solar energy heating system	Solar energy air conditioning system
I Zone with abundant resources	≥60	≥50	≥45
II Zone with relatively abundant resources	≥50	≥35	≥30
III Zone with general resources	≥40	≥30	≥25
IV Zone with poor resources	≥30	≥25	≥20

7.2.5 Auxiliary heat source of solar thermal utilization system shall be selected according to building service characteristics, heat consumption, energy supply, maintenance and management, and sanitation, and low-grade energy such as waste heat and residual heat and other renewable energy such as biomass and terrestrial heat is suggested to be utilized.

7.2.6 Solar collector and photovoltaic components shall be so set as to avoid being sheltered by themselves or main body of building. Sunshine duration of solar collector on day-lighting face in winter solstice shall not be less than 4h and that of photovoltaic components is not suggested to be less than 3h.

7.3 Ground Source Heat Pump System

7.3.1 When designing ground source heat pump system of public buildings, full-year dynamic load and heat supply and heat release of the system shall be calculated and analyzed, geothermal exchange system shall be determined, and composite thermal exchange system should be adopted.

7.3.2 High energy efficient water source heat pump unit shall be selected for design of ground source heat pump system, and energy saving measures is suggested to be taken to reduce transmission energy consumption of circulating water pump and improve energy efficiency of ground source heat pump system.

7.3.3 Performance of water source heat pump unit shall meet the requirements for operation parameters of geothermal exchange system, and equipment for heating and cooling of terminal shall be matched with operation parameters of water source heat pump unit.

7.3.4 Partial or whole heat recovery type water source heat pump unit is suggested to be adopted for public buildings with stable demand for hot water according to load characteristics. In case of full-year hot water supply, whole heat recovery type water source heat pump unit or water source hot water unit shall be selected.

Appendix A Calculation of Mean Heat Transfer Coefficient of External Walls

A. 0. 1 Mean heat transfer coefficient of external walls shall be calculated according to the requirements of the current national standard of GB 50176 *Thermal Design Code for Civil Building*.

A. 0. 2 For general buildings, the mean heat transfer coefficient of external walls may be also calculated according to following formula:

$$K = \varphi K_P \quad (A.0.2)$$

Where, K——the mean heat transfer coefficient of external walls [W/(m² · K)];

K_P——the heat transfer coefficient of main body of external walls [W/(m² · K)];

φ——the correction coefficient of heat transfer coefficient of main body of external walls.

A. 0. 3 The correction coefficient φ of heat transfer coefficient of main body of external walls may be selected according to Table A. 0. 3.

Table A. 0. 3 Correction Coefficient of Heat Transfer Coefficient of Main Body of External Walls φ

Climatic zoning	External insulation	Cavity insulation (self-insulation)	Internal insulation
Severe cold zone	1.30	—	—
Cold zone	1.20	1.25	—
Hot summer and cold winter zone	1.10	1.20	1.20
Hot summer and warm winter zone	1.00	1.05	1.05

Appendix B Building Envelope Thermal Performance Trade-off

B. 0. 1 Specialized computing software capable of automatically generating reference building calculation model meeting the requirements of this standard shall be adopted for building envelope thermal performance trade-off, and it shall have the following functions:

  1 Calculating hourly load over full-year 8760h;

  2 Hourly setting the number of indoor personnel, lighting power, equipment power, indoor temperature, and runtime of heating and air conditioning system during working days and festivals and holidays;

  3 Considering heat storage capacity of building envelope;

  4 Calculating more than 10 zones in building;

  5 Directly generating calculation report of building envelope thermal performance trade-off.

B. 0. 2 For building envelope thermal performance trade-off, energy consumption judgment shall be based on the general power consumption of heating and air conditioning in reference building and designed building. Coal and gas consumption for heating in reference building and designed building shall be converted into power consumption.

B. 0. 3 Energy consumption for air conditioning and heating in reference building and designed building shall be calculated by the same software, and data in typical meteorological year shall be adopted as meteorological parameters.

B. 0. 4 Accumulated full-year cold consumption and heat consumption in designed building shall be calculated in accordance with the following requirements:

  1 Building shape, size, orientation, internal space division and use function, building structure dimension, heat transfer coefficient of building envelope, construction, solar heat gain coefficient of external window (including transparent curtain wall), window to wall ratio, and roofing operable window area shall comply with those specified in design document.

  2 Air conditioning and heating of buildings shall be set according to full-year operated two-pipe fan coil systems. Building function zones shall be calculated based on the heating and air conditioning, except that zones are defined as non-air-conditioned zones in design document;

  3 Runtime of air conditioning and heating system, indoor temperature, lighting power density and switch time, room utilization area per capita and presence rate, fresh air rate for personnel and runtime of fresh air unit, power density and utilization rate of electrical equipment in building shall comply with Tables B. 0. 4-1~B. 0. 4-10.

Table B. 0. 4-1 Daily Runtime of Air Conditioning System and Heating System

Category	System working time	
Office building	Working days	7:00~18:00
	Festivals and holidays	—
Hotel building	All year around	1:00~24:00
Commercial building	All year around	8:00~21:00

Table B. 0. 4-1(continued)

Category	System working time	
Medical building-outpatient building	All year around	8:00~21:00
School building-teaching building	Working days	7:00~18:00
	Festivals and holidays	—

Table B. 0. 4-2 Indoor Temperature of Heating and Air-conditioning Zone (℃)

Building category	Runtime	Running mode	Indoor set temperature (℃) of heating and air conditioning zone at the following calculation hour (h)											
			1	2	3	4	5	6	7	8	9	10	11	12
Office and teaching buildings	Working days	Air conditioning	37	37	37	37	37	37	28	26	26	26	26	26
		Heating	5	5	5	5	5	12	18	20	20	20	20	20
	Festivals and holidays	Air conditioning	37	37	37	37	37	37	37	37	37	37	37	37
		Heating	5	5	5	5	5	5	5	5	5	5	5	5
Hotel building, inpatient department	All year around	Air conditioning	25	25	25	25	25	25	25	25	25	25	25	25
		Heating	22	22	22	22	22	22	22	22	22	22	22	22
Commercial building, outpatient building	All year around	Air conditioning	37	37	37	37	37	37	37	28	25	25	25	25
		Heating	5	5	5	5	5	5	12	16	18	18	18	18

Building category	Runtime	Running mode	Indoor set temperature (℃) of heating and air conditioning zone at the following calculation hour (h)											
			13	14	15	16	17	18	19	20	21	22	23	24
Office and teaching buildings	Working days	Air conditioning	26	26	26	26	26	26	37	37	37	37	37	37
		Heating	20	20	20	20	20	20	18	12	5	5	5	5
	Festivals and holidays	Air conditioning	37	37	37	37	37	37	37	37	37	37	37	37
		Heating	5	5	5	5	5	5	5	5	5	5	5	5
Hotel building, inpatient department	All year around	Air conditioning	25	25	25	25	25	25	25	25	25	25	25	25
		Heating	22	22	22	22	22	22	22	22	22	22	22	22
Commercial building, outpatient building	All year around	Air conditioning	25	25	25	25	25	25	25	25	37	37	37	37
		Heating	18	18	18	18	18	18	18	18	12	5	5	5

Table B. 0. 4-3 Lighting Power Density (W/m²)

Building category	Lighting power density
Office building	9.0
Hotel building	7.0
Commercial building	10.0
Medical building - outpatient building	9.0
School building - teaching building	9.0

Table B. 0. 4-4 Lighting Switch Time (%)

Building category	Runtime	Lighting switch time (%) in the following calculation hour (h)											
		1	2	3	4	5	6	7	8	9	10	11	12
Office and teaching buildings	Working days	0	0	0	0	0	0	10	50	95	95	95	80
	Festivals and holidays	0	0	0	0	0	0	0	0	0	0	0	0
Hotel building, inpatient department	All year around	10	10	10	10	10	10	30	30	30	30	30	30
Commercial building, outpatient building	All year around	10	10	10	10	10	10	10	50	60	60	60	60
Building category	Runtime	Lighting switch time (%) in the following calculation hour (h)											
		13	14	15	16	17	18	19	20	21	22	23	24
Office and teaching buildings	Working days	80	95	95	95	95	30	30	0	0	0	0	0
	Festivals and holidays	0	0	0	0	0	0	0	0	0	0	0	0
Hotel building, inpatient department	All year around	30	30	50	50	60	90	90	90	90	80	10	10
Commercial building, outpatient building	All year around	60	60	60	60	80	90	100	100	100	10	10	10

Table B. 0. 4-5 Building Area Per Capita in Different Types of Rooms (m²/person)

Building category	Building area per capita
Office building	10
Hotel building	25
Commercial building	8
Medical building - outpatient building	8
School building - teaching building	6

Table B. 0. 4-6 Hourly Presence Rate of Personnel in Room (%)

Building category	Runtime	Hourly presence rate (%) of personnel in room at the following calculation hour (h)											
		1	2	3	4	5	6	7	8	9	10	11	12
Office and teaching buildings	Working days	0	0	0	0	0	0	10	50	95	95	95	80
	Festivals and holidays	0	0	0	0	0	0	0	0	0	0	0	0
Hotel building, inpatient department	All year around	70	70	70	70	70	70	70	70	50	50	50	50
	All year around	95	95	95	95	95	95	95	95	95	95	95	95
Commercial building, outpatient building	All year around	0	0	0	0	0	0	0	20	50	80	80	80
	All year around	0	0	0	0	0	0	0	20	50	95	80	40
Building category	Runtime	Hourly presence rate (%) of personnel in room at the following calculation hour (h)											
		13	14	15	16	17	18	19	20	21	22	23	24
Office and teaching buildings	Working days	80	95	95	95	95	30	30	0	0	0	0	0
	Festivals and holidays	0	0	0	0	0	0	0	0	0	0	0	0
Hotel building, inpatient department	All year around	50	50	50	50	50	50	70	70	70	70	70	70
	All year around	95	95	95	95	95	95	95	95	95	95	95	95
Commercial building, outpatient building	All year around	80	80	80	80	80	80	80	70	50	0	0	0
	All year around	20	50	60	60	20	20	0	0	0	0	0	0

Table B.0.4-7 Fresh Air Rate Per Capita in Different Rooms [$m^3/(h \cdot person)$]

Building type	Fresh air rate
Office building	30
Hotel building	30
Commercial building	30
Medical building - outpatient building	30
School building - teaching building	30

Table B.0.4-8 Fresh Air Operation Condition (1 Represents Opening of Fresh Air, 0 Represents Closing of Fresh Air)

Building type	Runtime	\multicolumn{12}{c}{Fresh air operation condition at the following calculation hour (h)}											
		1	2	3	4	5	6	7	8	9	10	11	12
Office and teaching buildings	Working days	0	0	0	0	0	0	1	1	1	1	1	1
	Festivals and holidays	0	0	0	0	0	0	0	0	0	0	0	0
Hotel building,	All year around	1	1	1	1	1	1	1	1	1	1	1	1
inpatient department	All year around	1	1	1	1	1	1	1	1	1	1	1	1
Commercial building,	All year around	0	0	0	0	0	0	0	1	1	1	1	1
outpatient building	All year around	0	0	0	0	0	0	0	1	1	1	1	1

Building type	Runtime	\multicolumn{12}{c}{Fresh air operation condition at the following calculation hour (h)}											
		13	14	15	16	17	18	19	20	21	22	23	24
Office and teaching buildings	Working days	1	1	1	1	1	1	1	0	0	0	0	0
	Festivals and holidays	0	0	0	0	0	0	0	0	0	0	0	0
Hotel building,	All year around	1	1	1	1	1	1	1	1	1	1	1	1
inpatient department	All year around	1	1	1	1	1	1	1	1	1	1	1	1
Commercial building,	All year around	1	1	1	1	1	1	1	1	1	0	0	0
outpatient building	All year around	1	1	1	1	1	1	0	0	0	0	0	0

Table B.0.4-9 Power Density of Electrical Equipment in Different Rooms (W/m^2)

Building type	Power of electrical equipment
Office building	15
Hotel building	15
Commercial building	13
Medical building - outpatient building	20
School building - teaching building	5

Table B.0.4-10 Hourly Utilization Rate of Electrical Equipment (%)

Building type	Runtime	\multicolumn{12}{c}{Hourly utilization rate of electrical equipment at the following calculation hour (h)}											
		1	2	3	4	5	6	7	8	9	10	11	12
Office and teaching buildings	Working days	0	0	0	0	0	0	10	50	95	95	95	50
	Festivals and holidays	0	0	0	0	0	0	0	0	0	0	0	0
Hotel building,	All year around	0	0	0	0	0	0	0	0	0	0	0	0
inpatient department	All year around	95	95	95	95	95	95	95	95	95	95	95	95
Commercial building,	All year around	0	0	0	0	0	0	30	50	80	80	80	
outpatient building	All year around	0	0	0	0	0	0	20	50	95	80	40	

Table B. 0. 4-10(continued)

Building type	Runtime	Hourly utilization rate of electrical equipment at the following calculation hour (h)											
		13	14	15	16	17	18	19	20	21	22	23	24
Office and teaching buildings	Working days	50	95	95	95	95	30	30	0	0	0	0	0
	Festivals and holidays	0	0	0	0	0	0	0	0	0	0	0	0
Hotel building,	All year around	0	0	0	0	0	80	80	80	80	80	0	0
inpatient department	All year around	95	95	95	95	95	95	95	95	95	95	95	95
Commercial building,	All year around	80	80	80	80	80	80	80	70	50	0	0	0
outpatient building	All year around	20	50	60	60	20	20	0	0	0	0	0	0

B. 0. 5 Accumulated full-year cold consumption and heat consumption for reference building shall be calculated in accordance with the following requirements:

1 Building shape, size, orientation, internal space division and use function, and building structure dimension shall comply with those of designed building;

2 Building envelope shall be constructed as specified in building design document and building envelope thermal performance parameter shall meet the requirement of 3. 3 in this standard;

3 Runtime of air conditioning and heating system, indoor temperature, lighting power density and switch time, room utilization area per capita and presence rate, fresh air rate for personnel and runtime of fresh air unit, power density and utilization rate of electrical equipment in building shall comply with those specified in designed building;

4 Full-year operated two-pipe fan coil system shall be adopted for air conditioning and heating of buildings. Heating and air-conditioned zones shall be set as specified in designed building.

B. 0. 6 In calculating general power consumption of full-year heating and air conditioning of designed building and reference building, electrically driven water chilling unit shall be adopted for cooling source of air conditioning system; coal-fired boiler shall be adopted as heating source of heating system in severe cold zone and cold zone; gas-fired boiler shall be adopted as heating source of heating system in hot summer and cold winter zone, hot summer and warm winter zone, and temperate zone, and shall meet the following requirements:

1 General power consumption of full-year heating and air conditioning shall be calculated according to the following formula:

$$E = E_H + E_C \tag{B. 0. 6-1}$$

Where, E——the general power consumption of full-year heating and air conditioning (kWh/m^2);

E_C——the power consumption of full-year air conditioning (kWh/m^2);

E_H——the power consumption of full-year heating (kWh/m^2).

2 Power consumption of full-year air conditioning shall be calculated according to the following formula:

$$E_C = \frac{Q_C}{A \times SCOP_T} \tag{B. 0. 6-2}$$

Where, Q_C——the accumulated full-year cold consumption (calculated by dynamic simulation software) (kWh);

A——the total building area (m^2);

$SCOP_T$——the coefficient of comprehensive performance of cooling system, and 2. 50 is taken.

3 Power consumption of full-year heating in severe cold zone and cold zone shall be calculated

according to the following formula:

$$E_H = \frac{Q_H}{A\eta_1 q_1 q_2} \qquad (B.0.6\text{-}3)$$

Where, Q_H ——the accumulated full-year heat consumption (calculated by dynamic simulation software) (kWh);

η_1 ——the overall efficiency of heating system with coal-fired boiler acted as heating source, and 0.60 is taken;

q_1 ——the heat value of standard coal, and 8.14 kWh/kgce is taken;

q_2 ——the coal consumption for power generation (kgce/kWh), and 0.360kgce/kWh is taken.

4 Power consumption of full-year heating in hot-summer and cold-winter zone, hot-summer and warm-winter zone and temperate zone shall be calculated according to the following formula:

$$E_H = \frac{Q_H}{A\eta_2 q_3 q_2}\varphi \qquad (B.0.6\text{-}4)$$

Where, η_2 ——the overall efficiency of heating system with gas-fired boiler acted as heating source, and 0.75 is taken;

q_3 ——the heat value of standard natural gas, and 9.87 kWh/m³ is taken;

φ ——the conversion coefficient of natural gas and standard coal, and 1.21 kgce/m³ is taken.

Appendix C Building Envelope Thermal Performance Compliance Form

Table C Audit List for Building Envelope Thermal Performance Trade-off

Project name						
Project site						
Design organization						
Design date:				Climate zone		
Software adopted				Software version		
Building area		m²		Building appearance area		m²
Building volume		m²		Shape factor		

Window to wall ratio of designed building				The ratio of roof transparent part to total roof area M	Limit of M	
Facade 1	Facade 2	Facade 3	Facade 4			
					20%	

Building envelope parts	Designed building		Reference building		Whether meet the limit specified in this standard
	Heat transfer coefficient K W/(m²·K)	Solar heat gain coefficient SHGC	Heat transfer coefficient K W/(m²·K)	Solar heat gain coefficient SHGC	
Roof transparent part					
External window of facade 1 (including transparent curtain wall)					
External window of facade 2 (including transparent curtain wall)					
External window of facade 3 (including transparent curtain wall)					
External window of facade 4 (including transparent curtain wall)					

Building envelope parts	Designed building		Reference building		Whether meet the limit specified in this standard
	Heat transfer coefficient K W/(m²·K)	Solar heat gain coefficient SHGC	Heat transfer coefficient K W/(m²·K)	Solar heat gain coefficient SHGC	
Roofing		—		—	
External wall (including non-transparent curtain wall)		—		—	
Overhead or overhanging floor slab with bottom surface contacting outdoor air		—		—	
Partition and floor slab between non-heating room and heating room		—		—	

Table C(continued)

Building envelope parts	Designed building		Reference building		Whether meet the limit specified in this standard
	Thermal resistance of insulation material layer $R[(m^2 \cdot K)/W]$		Thermal resistance of insulation material layer $R[(m^2 \cdot K)/W]$		
Surrounding ground					
External wall of heating basement in contact with soil					
Deformation joint (in the case of internal insulation of two side walls)					
Judgment of basic requirements for trade-off	Basic requirement of heat transfer coefficient of building envelope $K[W/(m^2 \cdot K)]$		Whether designed building meets the basic requirement		
	Roofing				
	External wall (including non-transparent curtain wall)				
	External window (including transparent curtain wall)				
	Solar heat gain coefficient SHGC				
	Whether building envelope meets basic requirements		Yes/No		
Trade-off result	Designed building (kWh/m²)		Reference building (kWh/m²)		
General power consumption for full-year heating and air conditioning					
Trade-off conclusion	Is designed building envelope thermal performance qualified /unqualified?				

Appendix D Insulation Thickness of Pipes, Ducts and Equipments

D.0.1 Economical thermal insulation thickness of heating pipe may be selected according to those specified in Table D.0.1-1~Table D.0.1-3. Thermal insulation thickness of heating equipment may be increased by 5mm according to thermal insulation thickness of the maximum caliber pipe.

Table D.0.1-1 Economical Thermal Insulation Thickness for Flexible Foamed Rubber
Plastics of Indoor Heating Pipe (Caloric Value 85 Yuan/GJ)

Max medium temperature (℃)	Thermal insulation thickness (mm)						
	25	28	32	36	40	45	50
60	≤DN20	DN25~DN40	DN50~DN125	DN150~DN400	≥DN450	—	—
80	—	—	≤DN32	DN40~DN70	DN80~DN125	DN150~DN450	≥DN500

Table D.0.1-2 Economical Thermal Insulation Thickness for Centrifugal Glass
Cotton of Heating Pipe (Caloric Value 35 Yuan/GJ)

	Max medium temperature (℃)	Thermal insulation thickness (mm)								
		25	30	35	40	50	60	70	80	90
Indoor	60	≤DN40	DN50~DN125	DN150~DN1000	≥DN1100	—	—	—	—	—
	80	—	≤DN32	DN40~DN80	DN100~DN250	≥DN300	—	—	—	—
	95	—	—	≤DN40	DN50~DN100	DN125~DN1000	≥DN1100	—	—	—
	140	—	—	—	≤DN25	DN32~DN80	DN100~DN300	≥DN350	—	—
	190	—	—	—	—	≤DN32	DN40~DN80	DN100~DN200	DN250~DN900	≥DN1000
Outdoor	60	—	≤DN40	DN50~DN100	DN125~DN450	≥DN500	—	—	—	—
	80	—	—	≤DN40	DN50~DN100	DN125~DN1700	≥DN1800	—	—	—
	95	—	—	≤DN25	DN32~DN50	DN70~DN250	≥DN300	—	—	—
	140	—	—	—	≤DN20	DN25~DN70	DN80~DN200	DN250~DN1000	≥DN1100	—
	190	—	—	—	—	≤DN25	DN32~DN70	DN80~DN150	DN200~DN500	≥DN600

Table D.0.1-3 Economical Thermal Insulation Thickness for Centrifugal Glass Cotton of Heating Pipe (Caloric Value 85 Yuan/GJ)

Max medium temperature (℃)		Thermal insulation thickness (mm)								
		40	50	60	70	80	90	100	120	140
Indoor	60	≤DN50	DN70~DN300	≥DN350	—	—	—	—	—	—
	80	≤DN20	DN25~DN70	DN80~DN200	≥DN250	—	—	—	—	—
	95	—	≤DN40	DN50~DN100	DN125~DN300	DN350~DN2500	≥DN3000	—	—	—
	140	—	—	≤DN32	DN40~DN70	DN80~DN150	DN200~DN300	DN350~DN900	≥DN1000	—
	190	—	—	—	≤DN32	DN40~DN50	DN70~DN100	DN125~DN150	DN200~DN700	≥DN800
Outdoor	60	—	≤DN80	DN100~DN250	≥DN300	—	—	—	—	—
	80	—	≤DN40	DN50~DN100	DN125~DN250	DN300~DN1500	≥DN2000	—	—	—
	95	—	≤DN25	DN32~DN70	DN80~DN150	DN200~DN400	DN500~DN2000	≥DN2500	—	—
	140	—	—	≤DN25	DN32~DN50	DN70~DN100	DN125~DN200	DN250~DN450	≥DN500	—
	190	—	—	—	≤DN25	DN32~DN50	DN70~DN80	DN100~DN150	DN200~DN450	≥DN500

D.0.2 Minimum thermal insulation thickness of chilled water pipe for indoor air conditioning may be selected according to those specified in Table D.0.2-1~Table D.0.2-2; Insulation thickness of cold storage equipment may be increased by 5mm~10mm according to cold insulation thickness of the maximum caliber pipe at corresponding medium temperature.

Table D.0.2-1 Minimum Thermal Insulation Thickness of Chilled Water Pipe for Indoor Air Conditioning (Medium Temperature ≥5℃) (mm)

Area	Flexible foamed rubber plastics		Glass cotton tube	
	Pipe diameter	Thickness	Pipe diameter	Thickness
Relatively arid area	≤DN40	19	≤DN32	25
	DN50~DN150	22	DN40~DN100	30
	≥DN200	25	DN125~DN900	35
Relatively humid area	≤DN25	25	≤DN25	25
	DN32~DN50	28	DN32~DN80	30
	DN70~DN150	32	DN100~DN400	35
	≥DN200	36	≥DN450	40

Table D.0.2-2 Minimum Thermal Insulation Thickness of Chilled Water Pipe for Indoor Air Conditioning (Medium Temperature ≥ −10℃) (mm)

Area	Flexible foamed rubber plastics		Polyurethane foam	
	Pipe diameter	Thickness	Pipe diameter	Thickness
Relatively arid area	≤DN32	28	≤DN32	25
	DN40~DN80	32	DN40~DN150	30
	DN100~DN200	36	≥DN200	35
	≥DN250	40	—	—
Relatively humid area	≤DN50	40	≤DN50	35
	DN70~DN100	45	DN70~DN125	40
	DN125~DN250	50	DN150~DN500	45
	DN300~DN2000	55	≥DN600	50
	≥DN2100	60	—	—

D.0.3 Economical thermal insulation thickness of Indoor domestic hot water pipe may be selected according to those specified in Table D.0.3-1 and Table D.0.3-2.

Table D.0.3-1 Economical Thermal Insulation Thickness of Indoor Domestic Hot Water Pipe (Indoor 5℃ Full-year ≤105 days)

Medium temperature \ Thermal insulation material	Centrifugal glass cotton		Flexible foamed rubber plastics	
	Nominal pipe diameter (mm)	Thickness (mm)	Nominal pipe diameter (mm)	Thickness (mm)
≤70℃	≤DN25	40	≤DN40	32
	DN32~DN80	50	DN50~DN80	36
	DN100~DN350	60	DN100~DN150	40
	≥DN400	70	≥DN200	45

Table D.0.3-2 Economical Thermal Insulation Thickness of Indoor Domestic Hot Water Pipe (Indoor 5℃ Full-year ≤150 days)

Medium temperature \ Thermal insulation material	Centrifugal glass cotton		Flexible foamed rubber plastics	
	Nominal pipe diameter (mm)	Thickness (mm)	Nominal pipe diameter (mm)	Thickness (mm)
≤70℃	≤DN40	50	≤DN50	40
	DN50~DN100	60	DN70~DN125	45
	DN125~DN300	70	DN150~DN300	50
	≥DN350	80	≥DN350	55

D.0.4 Minimum thermal resistance of air duct insulation layer for indoor air conditioning may be selected according to those specified in Table D.0.4.

Table D.0.4 Minimum Thermal Resistance of Air Duct Insulation Layer for Indoor Air Conditioning

Type of air duct	Applicable medium temperature (℃)		Minimum thermal resistance $R[(m^2 \cdot K)/W]$
	Lowest temperature of cooling medium	Highest temperature of heating medium	
General air conditioning duct	15	30	0.81
Low temperature duct	6	39	1.14

Explanation of Wording in This Standard

1 Words used for different degrees of strictness are explained as follows in order to mark the differences in executing the requirements in this standard:

1) Words denoting a very strict or mandatory requirement:

   "must" is used for affirmation, "must not" for negation;

2) Words denoting a strict requirement under normal conditions:

   "shall" is used for affirmation, "shall not" for negation;

3) Words denoting a permission of a slight choice or an indication of the most suitable choice when conditions permit:

   "be suggested" is used for affirmation, "be not suggested" for negation;

4) "May" is used to express the option available, sometimes with the conditional permit.

2 "Shall meet the requirements of..." or "Shall be carried out according to..." is used in this standard to indicate that it is necessary to comply with the requirements stipulated in other relative standards.

List of Quoted Standards

1 GB 50015 *Code for Design of Building Water Supply and Drainage*
2 GB 50034 *Standard for Lighting Design of Buildings*
3 GB 50176 *Thermal Design Code for Civil Building*
4 GB 50314 *Standard for Design of Intelligent Building*
5 GB 50555 *Standard for Water Saving Design in Civil Building*
6 GB 50736 *Design Code for Heating Ventilation and Air Conditioning of Civil Buildings*
7 GB/T 7106 *Gradations and Test Methods of Air Permeability, Watertightness, Wind Load Resistance Performance for Building External Windows and Doors*
8 GB/T 8175 *Guide for Design of Thermal Insulation of Equipments and Pipes*
9 GB/T 14295 *Air Filters*
10 GB 19762 *Limited Values of Energy Efficiency and Evaluating Values of Energy Conservation of Centrifugal Pump for Fresh Water*
11 GB 20052 *Minimum Allowable Values of Energy Efficiency and Energy Efficiency Grades for Three-phase Distribution Transformers*
12 GB/T 21086 *Building Curtain Walls*
13 JGJ/T 163 *Code for Lighting Design of Urban Nightscape*
14 CJ/T 164 *Domestic Water Saving Devices*